It takes a sort of cosmic chutzpah to announce that your book is going to explain the purpose of the universe, and that predisposed me in its favour before I even began to read the text. But what I enjoyed most was the author's calm and level-headed approach to that fundamental question: Why? I suppose it's the question most young people begin with when they start to develop intellectual curiosity, as well as being the source of every system of religion and of science as well. Philip Goff explains that 'purpose' is not some emotional human need with little ultimate importance, but a quality that seems to be built into the very nature of things. It's nothing less than thrilling to follow his argument, and to regain that sense of connectedness that's so important not just to our well-being, but to our very survival.

Philip Pullman

This book is a tour de force. If you have ever wondered what the point is in living, whether the universe itself has any purpose (and if it does, whether that is best explained by the existence of God), why the universe exists at all for that matter, what the nature of consciousness is and how it fits into the universe as a whole, this is the book for you. Written in an engaging and easy to follow style, Goff presents a highly original, unified, and thought-provoking world view. It is rare to read anything that makes one seriously question one's basic assumptions about reality. Goff's book does just that. The result is something head spinning. I cannot recommend it highly enough.

Michael Tye Professor of Philosophy,
The University of Texas at Austin

Am I here by accident? Is there a purpose? This is contentious territory in science and philosophy. Goff offers a lucid and riveting account of key ideas, data, and theories. He then, with a rare audacity, blazes new trails. It is fascinating terrain to explore, and Goff proves an expert and genial guide.

Donald Hoffman, Professor of Cognitive Science,
University of California, Irvine

The best metaphysical pictures help us fulfill three aims: give us a sense of the world and how we as human beings fit within it, provide an ethical guide, and help us make some peace with our condition. Philip Goff's brilliant new book *Why? The Purpose of the Universe* does all three by making a compelling case for teleological cosmopsychism. With this unique position that is neither God nor atheism, Goff gives us a glimpse of the immense creative potential of

Why?

the universe. He outlines an attractive picture of spiritual belonging and practice in a godless world that is far from nihilistic.

Helen De Cruz, Danforth Chair in the Humanities,
Saint Louis University

It might sound surprising, but the progress of physics, astronomy and cosmology in recent decades has raised important questions about the meaning and purpose in the universe, and of the universe. "Nature has been kinder to us than we had any right to expect", wrote Freeman Dyson in 1971, "it almost seems as if the universe must in some sense have known that we were coming". So, what now? Philip Goff has provided a discussion of these important issues that is informed, accessible, original and entertaining. This is a book worth reading, and worth thinking hard about.

Luke Barnes, Lecturer in Astronomy and Cosmology,
Western Sydney University

Why? is a terrific book. For a work in philosophy, it is unusually fun to read. Goff clearly has a gift for making rigorous philosophy accessible to a broad audience. Part of his genius is the way he weaves his main arguments into a narrative about his own intellectual journey.

Paul Draper, Professor of Philosophy,
Purdue University

Why? is simultaneously accessible and profound, comprehensible to the general reader and full of novel ideas sure to challenge professional philosophers. Goff offers an intriguingly weird vision of the cosmos, neither atheistic nor orthodox, pushing beyond the boundaries of both ordinary scientific thinking and ordinary religious apologetics.

Eric Schwitzgebel, Professor of Philosophy,
University of California, Riverside

A brilliant book! Goff takes us to the edges of physics and philosophy to make a compelling case for cosmic purpose. The presentation is clear, innovative, and provocative. True to form, Goff's ideas are not anchored to convention or tradition, but he instead lights a torch on an original path of discovery. I came away feeling that Goff's work contributes to the purpose of the universe in a profound and beautiful way. I highly recommend this book to anyone interested in the big question of why we are here and what life might be about.

Josh Rasmussen, Associate Professor of Philosophy,
Azusa Pacific University

What's the meaning of life? Raw science tells us we live in an uncaring universe, devoid of purpose and oblivious to the wants and needs of humans. Or does it? In this new book, Goff explores purpose in the cosmos, not a purpose rooted in religion, but in a conscious fabric of the universe itself. Whilst Goff and I have argued over the implications of cosmological fine-tuning and the statistics of typing monkeys, the contents will certainly make you think about life and its meaning.

Geraint Lewis, Professor of Astrophysics,
Sydney Institute for Astronomy, The University of Sydney

Why?

The Purpose of the Universe

PHILIP GOFF

OXFORD
UNIVERSITY PRESS

OXFORD
UNIVERSITY PRESS

Great Clarendon Street, Oxford, OX2 6DP,
United Kingdom

Oxford University Press is a department of the University of Oxford.
It furthers the University's objective of excellence in research, scholarship,
and education by publishing worldwide. Oxford is a registered trade mark of
Oxford University Press in the UK and in certain other countries

Published in the United States of America by Oxford University Press
198 Madison Avenue, New York, NY 10016, United States of America

British Library Cataloguing in Publication Data
Data available

Library of Congress Control Number: 2023934669

ISBN 978-0-19-888376-0

DOI: 10.1093/oso/9780198883760.001.0001

Printed by Integrated Books International, United States of America

For Hannah and Matilda, a crucial part of the purpose of my existence

Contents

How to Read This Book

Academic philosophers tend to talk to themselves. They write complicated, jargon-filled books that are inaccessible to anyone who doesn't have a PhD in philosophy. I've written one of them myself, so I should know. I wanted this book to be *both* a significant contribution to philosophy *and* accessible to a broader audience. It's a hard act to pull off. Academic philosophers like myself are trained to construct watertight arguments, which means covering all possible objections. This leaves me with a difficult dilemma. On the one hand, if I don't cover all objections, other philosophers will think my arguments are rubbish. On the other hand, it's hard to cover all objections without making the book dense and inaccessible to the general reader.

I've tried to cover both bases by dividing most of chapters into a more accessible first section, followed by a 'Digging Deeper' second section. Sometimes the 'Digging Deeper' section consists of responses to objections, other times it delves more deeply into the topic. In either case, those wanting the big picture without some of the challenging details, at least on the first read through, can avoid the 'Digging Deeper' sections without loss of continuity. Perhaps you might want to return to some of the complexities after reaching the end, to get a deeper understanding of the claims and arguments of the book. Alternately, you might want to flick through the 'Digging Deeper' sections as you go along to see if your burning objection is covered, or if there's something of particular interest that you'd like to read more about.

Even then, the book might be challenging. Whoever said contemplating the ultimate purpose of existence would be easy? Some chapters are harder than others: the chapters focusing on consciousness (3 and 6) deal with inherently difficult concepts and so are perhaps the most challenging, the chapters focusing on the implications for human existence (1, at least prior to the 'Digging Deeper' section, and 7) are probably the most accessible. I would encourage readers to have a go at all of it, without worrying too much if you don't get some bits. If you're struggling, you can skip or skim Chapters 3 and 6 without losing the central argument of the book. Overall, I'm confident the effort to engage with the deep mystery of existence will yield big returns.

Note to academic philosophers: This book isn't structured like a standard philosophical text. Chapters don't kick off by laying out exactly what's going to happen when, where, and why. There are personal stories and digressions, and sometimes I move between different points without clearly signalling the transition. Where the argument is beginning to get too dense at the cost of accessibility, I've shifted some of it to endnotes. All of this is necessary in a book aimed at a general audience, but may require a bit of work—not too much I hope—on the part of the academic philosopher in piecing together the argument from the surrounding narrative. Enjoy.

1

What's the Point of Living?

One day all of the energy in the universe will be used up. The stars will all go out. All matter will be swallowed up by black holes, black holes that will eventually evaporate away. This will be the beginning of an infinitely long epoch in which all activity and interaction has ceased. There will never again be intelligent life, anywhere, for ever more.[1]

Christian philosopher William Lane Craig argues that this is the fate we are resigned to without God and that it is a fate that renders our lives meaningless.[2] The 11-year-old Alvy Singer in the movie *Annie Hall* refused to do his homework in the light of the eventual death of the universe ('What's the point?'). Craig even goes so far as to say that killing and loving are morally equivalent if the universe is eventually going to come to nothing and hence our lives have no ultimate significance.

It's not only religious believers who worry about these apparent implications of atheism. Atheist philosopher David Benatar argues that while our lives may seem to have meaning and significance from the perspective of somebody engaged in the hustle and bustle of life, when we zoom out to a cosmic perspective, we can see that our lives have very little significance in the broader scheme of things.[3] In his book *The Human Predicament*, Benatar offers an analogy of small children engaged in a heated debate, perhaps over whose turn it is to play with the ball. From the perspective of those children, who has the ball is of the utmost importance and significance. But if we zoom out to a slightly broader perspective, we can see that the focus of their quarrel has little significance. In the same way, the concerns of adults—our hopes and fears, our plans for life—are insignificant from the perspective of the universe as a whole, and certainly not worth the concern we invest in them.

What, then, is the point of living? Benatar draws a grim conclusion from his views on our insignificance: it would have been better if we'd never existed. He does not advocate suicide; now we're here, we might as well carry on. But he does argue that it is morally wrong to create new children who will grow up to live lives without significance. The morally correct things to do, according to Benatar, is to let the human race pass peacefully out of existence.

This is the position known as *anti-natalism*, a view which has almost become a kind of religion in its own right, at least to the extent that its small but dedicated adherents find comfort and even purpose in its tenets. In February 2019, Raphael Samuel, a 27-year-old Indian, announced to the world that he was suing his parents for bringing him into existence. Samuel had had a happy childhood and was close to his parents, but nonetheless resented being brought into existence without his consent. He told the BBC: 'It was not our decision to be born. Human existence is totally pointless.'[4]

It is perhaps not so surprising that people should be asking whether it's right to have children in a world threatened by imminent climate break-down. In my own case, I remember experiencing some guilt after my first child was born. We lived in Budapest at the time, and I have a vivid memory of carrying my baby from the car down the snowy street (she was born just before Christmas) to our apartment. As I passed the shop on the corner, the owner, who happened to be standing outside, smiled and said 'Gratulálunk!' At that moment, I was surprised to suddenly feel a kind of self-consciousness, almost embarrassment. I had been walking on cloud nine since the birth of my child twenty-four hours ago. But was this joy just for my sake, and for that of my partner? This child hadn't chosen to be born into a troubled world, and now, being 'caught in the act' of bringing her home to start her life, I found myself asking a question which had never occurred to me pre-viously: Was it fair to force this beautiful baby to live in an ugly world?

My concerns were rooted in the current state of the world. This is not the motivation of the anti-natalists, however. Their ethical position is rooted not in our contingent political circumstances but in the inherently unsatis-factory nature of human existence. As Benatar puts it 'Every birth is a death in waiting.'[5] Even if we one day achieve a Utopia of freedom, equality, and harmony with the natural environment, Benatar argues that life would still be insignificant from the cosmic perspective, and voluntary extinction the only humane option.

Benatar's concern about our cosmic insignificance is rooted in his belief that we live in a meaningless universe. But what if we don't live in a mean-ingless universe? What if the universe we live in has a purpose? Pierre Teilhard de Chardin was a palaeontologist and heretical Catholic priest, who lived in the late nineteenth and early 20th century.[6] At a time when the church was anxious about Darwinism, Teilhard was inspired by the idea of evolution. Whereas Darwin saw evolution in biology, Teilhard believed that the whole universe was evolving towards a higher state. Teilhard's argument was partly inductive. He looked to the past and saw the emergence of phenomena that, in his view, could not be accounted for in terms of what had gone before: life, consciousness, reason, moral awareness. The emergence of these phenomena seemed to Teilhard to point to a trajectory to ever greater states of existence, and he anticipated that this trajectory would continue into the future.

Some have even argued that Teilhard predicted the internet, as he believed the next state of cosmic evolution would arise from human beings becoming more and more informationally connected up, leading ultimately to a new form of life and consciousness which he called the 'noosphere.' Whilst Teilhard believed that cosmic evolution was in some sense inevitable, our actions could hasten or delay its progress. The emergence of the noosphere depended upon humankind being in a spiritually advanced enough state to move to that new stage of cosmic evolution. By choosing to build a world of peace and justice—rather than to sow hatred and division—and to raise our consciousness through meditation and simple living—rather than to indulge in wealth and excessive consumption—Teilhard believed that we can collectively contribute to the advancement of the universe, the end point of which would involve the physical universe becoming unified with the Divine. Quite a different picture from the cosmic destiny outlined in the first paragraph of this chapter!

Forget for a moment whether or not you're convinced by Teilhard's argument for cosmic purpose. Our first question is not whether cosmic purpose exists—there'll be plenty of time for that—but on what bearing the existence or non-existence of cosmic purpose has on the human condition. If Teilhard's view *were* true, Benatar's pessimistic evaluation of human existence would be transformed. Recall Benatar's concern is that, when we zoom out to the cosmic perspective, our lives turn out to have miniscule significance. But if Teilhard's visions of reality turned out to be right, this would be false. Our good actions would contribute, even in some small way, to the purposes of reality as a whole. Even when we zoom out to the cosmic perspective, our lives would have significance.

Teilhard was a Catholic believer, albeit of a rather unusual kind. As such, he believed in a personal God who was the ultimate instigator, and indeed the ultimate endpoint, of cosmic evolution. But we can imagine a hypothesis very much like Teilhard's but without a personal God. In fact, the 19th-century philosopher Samuel Alexander defended such a view.[7] For Alexander, the cosmic evolution of the universe was driven not by a personal God but by a natural tendency of the universe to move towards higher states of being, a drive Alexander called *Nisus*. What is important for the possibility of human life having cosmic significance is that the universe has a purpose, that it is directed towards some higher state of being, and that human beings can play a role in advancing the universe towards that higher state of being. Whether cosmic purpose is imposed on the universe externally by a supernatural god, or whether it arises from natural tendencies of the universe itself, is irrelevant.

Whether or not our lives have cosmic significance, then, does seem to depend on whether or not the universe has a purpose. But I'm sure many readers are by now bursting to shout out loud the following question (those, that is, that have not already done so): Who the hell cares about cosmic significance??!! Can't we live perfectly happy meaningful lives here on little old planet Earth, even if our lives lack cosmic significance?

For what it's worth, I agree that Craig and Benatar's concerns with cosmic significance are overblown. Even if our lives lack cosmic significance, they can still be filled with meaning in the here and now, provided we are able to engage in activities which are objectively meaningful, such as creativity, learning, and showing kindness to others. On the other hand, I think certain humanists who argue that cosmic significance would be totally irrelevant to human meaning are going too far in the other direction. We want our lives to have significance, we want to make a difference. If we were able to contribute, even in some small way, to the good purposes of *the whole of reality*, that would be about as big a difference as you can imagine making and would consequently greatly add to the meaning of our lives. A universe without cosmic purpose would not be as horrific as Craig and Benatar make out. But if it turned out there were a cosmic purpose, one that was good and that we could contribute to through our actions, that would be tantamount to winning the Reality Lottery.

I'd like to share some good news with you. I believe that there is overwhelming evidence for the existence of cosmic purpose. Whether it is of the kind that would add great meaning to our lives is a further question, and one we will return to in the final chapter. In the meantime, let us explore the case for cosmic purpose.

Choice Point: You now have the option of moving to the next chapter and starting to explore the case for cosmic purpose. However, if you want to dig a little deeper on this theme, I think there is a more subtle connection between human meaning and cosmic purpose, which we will explore in the 'Digging Deeper' section of this chapter.

Digging Deeper

Is Meaning in the Eye of the Beholder?

Why think my life is meaningful only if it's wrapped up in some grand cosmic drama? Even if we're alone in the universe, on one lonely planet with no cosmic significance, can't we nonetheless make our own meaning? If I find certain activities meaningful—whether it's finding a cure for cancer or just doing crossword puzzles—doesn't it follow that those activities *are* meaningful, at least for me? And if I'm lucky enough to spend a fair bit of time engaging in the activities I find meaningful, surely that means my life *is* meaningful. Isn't it up to me to decide what makes my life meaningful?

We can call the position I have just articulated *subjectivism* about meaning (with the caveat that the word 'subjectivism' is used in a bewildering variety of ways in philosophy).[8] For the meaning subjectivist, what gives my life meaning is determined by what I feel to be meaningful, or by what I value. If one person strongly values scientific research, then scientific research is meaningful for that person. If another person puts stamp collecting above all other life goals, then stamp collecting is deeply meaningful for that person. On this view, there are no externally given facts about which kinds of activities do or don't constitute a meaningful life. Meaning is in the eye of the beholder.

Subjectivist views in Western philosophy are often inspired by the work of the great enlightenment Scottish philosopher David Hume. Hume is a hero to many humanist philosophers, a kind of secular saint for his charming way of combining cautious skepticism with a cheery disposition.[9] Whilst Hume did not talk explicitly about the meaning of life, he held a subjectivist view of value in general.[10] For Hume, 'good' and 'bad,' 'right' and 'wrong,' are not features of the external world. The 'badness' of, say, a horrific murder, isn't to be found in the act itself, but in our negative feelings about the murder. Murder is bad because we feel it to be bad, and not vice versa. As Shakespeare put it,

there is nothing either good or bad, but thinking makes it so.

I decided when I was about 15 that I was a subjectivist about morality. I can't remember exactly what my reasoning was, but it just seemed to me obviously true that 'good' and 'bad' are in the eye of the beholder. What could it even mean for 'good' and 'bad' to be objective features of the world, in the way mass and charge are? I was not unusual in this regard. It is very common for philosophy undergraduates to think that moral subjectivism is 'obviously true' (these students are usually the loudest in the seminar group).

Philosophy for me isn't just an abstract exercise but something I live out. During my ethical subjectivist teenage years, I kissed my best friend John's girlfriend, telling him to his face that it was okay because 'good' and 'bad' depend on how you feel and it didn't feel bad. For his part, John decided on a creative response. With the help of a few friends, he kidnapped me, tied me up, and bleached my hair bright white. Even in the absence of moral objectivity, there are ways of stopping people doing shitty things.

By the time I began my graduate study at the University of Reading in my mid-20s, I'd learnt to behave in a more civilized manner, at least superficially. But I was still a moral subjectivist, convinced that the more empathy-based moral commitments I now had were still just expressions of my contingent emotions, the result of the way I happened to feel. But what I think has helped me most in my journey as a philosopher is that I'm insatiably curious about the views of people who disagree with me. The more they disagree with me, the more curious I am. I want to get inside their heads, to understand why they are thinking the way they do.

At the University of Reading, there were a couple of famous 'moral objectivists,' people who think there are objective facts about morality, in something like the way there are objective facts of science. For a moral objectivist, the fact that, say, the slave trade was wrong is as much an objective part of reality as the fact that the world is round. One of these moral objectivists was Jonathan Dancy, one of the few philosophers to have appeared on a primetime US chat show, on account of the fact that his daughter-in-law is Clare Danes, star of one of my favourite TV series *Homeland* (the interview with Dancy on *The Late Late Show With Craig Ferguson* is well worth a watch on YouTube). I craved to understand how on earth these distinguished professors, who had thought and read so much, could hold this bizarre position, and so I took every opportunity to interrogate them, both inside and outside the classroom. These were some of the most memorable conversations of my life. Whilst they didn't make me into a moral objectivist, they did persuade me that value subjectivism is ultimately an unsustainable position. I won't be able here to lay out the full story of all of the arguments that moved me, but I can lay out a couple of key moments in my intellectual journey.

The first thing that raised doubts about subjectivism was realizing that subjectivism about meaning has some pretty counterintuitive implications. Imagine a hypothetical person Susan (Susan is the star of all of my books) who has a single overriding goal that she bases her life around: counting blades of grass. Susan values counting blades of grass deeply. Her fundamental aim in life is to count as many blades of grass as she possibly can. She sweats, toils, cheats, and lies, all to the end of counting as many blades of grass as she can before she dies. Let's also add that Susan doesn't enjoy counting blades of grass. There is no incoherence here. People can dedicate their lives to activities they don't enjoy, if they value them enough. Many a creative is tortured by the creative process (indeed, writing a book can be fairly painful...). Because of her dedication, Susan manages to count a huge number of blades of grass during the span of her life.

If subjectivism about meaning is correct, we ought to say that Susan had a deeply meaningful life, given that she managed to spend a great deal of time engaged in an activity she valued highly.[11] I find this difficult to swallow. I can't shake the conviction that counting blades of grass seems like such a totally pointless waste of time. Contrast Susan's life with that of a doctor who works to reduce suffering, or a scientist who contributes to a deeper understanding of the universe, or just somebody who works hard to have a pleasant life for themselves and their family and friends. These all seem like worthwhile activities, and in contrast counting blades of grass seems like a tragic waste of a life.[12]

When I've debated this issue with people (I spend too much time arguing on Twitter...), a common response has been to suggest that human ingenuity can find meaning in even the most mundane activities. Maybe Susan has learnt the subtle variations among different blades of grass, and takes pleasure in cataloguing these differences. Maybe she has learnt to see the beauty in the patterns the blades make as the wind blows over them. But interpreting Susan's activities in this way is cheating. As I originally described the thought experiment, Susan's fundamental aim was to count blades of grass *for its own sake*, not for the sake of finding beauty or pleasure. Indeed, I specified that Susan does not enjoy counting blades of grass.[13]

Philosophers distinguish between 'intrinsic' and 'instrumental' goals. An instrumental goal is one pursued for the sake of some other goal. For example, most people work not for the sake of working, but for the sake of earning money. An intrinsic goal, in contrast, is one pursued for its own sake. Pleasure is a standard example. I seek the pleasure of the ice cream not for the sake of some other goal, but as an end in itself. As I'm imagining the Susan example, for her, counting blades of grass is not an instrumental goal but an intrinsic goal. Susan counts blades of grass for the sake of counting blades of grass.[14]

Another common objection to my Susan thought experiment is just to deny that anybody *could* have something so pointless as counting blades of grass as a fundamental life goal.[15] I kind of agree with that, but it effectively concedes the inadequacy of subjectivism. To say that an activity is so pointless that nobody would centre their lives around it is to appeal to some external, objective standard of which activities are or are not worth doing. But if subjectivism is true, there is no such external standard. People's arbitrary whims are the ultimate arbiter of what is worth doing.

I've also occasionally found people worrying there is an implicit authoritarianism in my condemnation of Susan's blade of grass counting, as though I'm envisaging a society where people's life goals are assessed by the 'Committee for the Assessment of Worthwhile Life Goals' for approval before they are allowed to live their lives as they choose. But this objection confuses moral and legal constraints. In supporting a free society, I approve of people's saying and doing whatever they want, so long as they don't hurt anyone (which incidentally implies ensuring that everyone has enough money, as money is a kind of freedom: the freedom to have and to do). This doesn't mean, however, that I think everybody is equally right! A free society doesn't require us all to approve of each other's choices. It means we allow people to make their own choices regardless of whether or not we approve. I don't want anyone to stop Susan counting her blades of grass. I just happen to think she's wasting her life.

It was these counterintuitive implications which started me on the road away from subjectivism.[16] What pushed me beyond the point of no return was grasping a deep tension at the heart of the David Hume-inspired moral philosophy I had adopted at this point.[17] To explain this, we need to dig a bit deeper into the technicalities of Hume's view of morality.

Hume is often described as thinking that you can't get an 'ought' from an 'is.' What is meant by this is that Hume proposed a kind of *logical gap* between statements that concern value and statements that don't concern value. More specifically, Hume claimed that it is impossible to move in reason from cold-blooded facts about the world to conclusions about what one *ought* to do. Hume put it thus:

> In every system of morality, which I have hitherto met with, I have always remarked, that the author proceeds for some time in the ordinary way of reasoning, and establishes the being of a God, or makes observations concerning human affairs; when of a sudden I am surprised to find, that

instead of the usual copulations of propositions, is, and is not, I meet with no proposition that is not connected with an ought, or an ought not. This change is imperceptible; but is, however, of the last consequence. For as this ought, or ought not, expresses some new relation or affirmation, it's necessary that it should be observed and explained; and at the same time that a reason should be given, for what seems altogether inconceivable, how this new relation can be a deduction from others, which are entirely different from it. But as authors do not commonly use this precaution, I shall presume to recommend it to the readers; and am persuaded, that this small attention would subvert all the vulgar systems of morality, and let us see, that the distinction of vice and virtue is not founded merely on the relations of objects, nor is perceived by reason.[18]

For the vast majority of us who feel empathy for others, suffering naturally motivates us to help. But for Hume, there is no rational inference to be made from the fact that someone is in terrible pain to the conclusion that anybody *ought* to help. Statements of fact and statements about what 'ought' to be done are just too different to allow such inferences to be drawn.

It was because of this logical gap between facts and values that Hume decided that morality must come from *feelings* rather than *reason*. If what ought to be done were logically entailed by the cold-blooded facts of science, then it would be the job of reason to sniff out these logical implications, and thereby discern the truths of morality. But in the absence of any such logical connections, it seems that reason is powerless to discern moral truths. And if the source of moral truth is not reason, thought Hume, then it must be our subjective feelings: things are good and bad because we feel them to be good and bad. Hume concluded on this basis that our fundamental goals in life are determined by our feelings, not by reason. The job of reason is merely to help us best achieve the goals our feelings set for us. He summed this up with his famous declaration that 'Reason is, and ought only to be the slave of the passions, and can never pretend to any office than to serve and obey them.'[19]

This all seemed to make perfect sense to me when I learnt it as an undergraduate. All that changed during my first year as a graduate student. I can still vividly remember the moment, sat in *The Queen's Head* pub in Reading, when I realized that there was a deep inconsistency at the core of the Hume-inspired subjectivist view that I had come to be so confident of. Appreciating this was one of the most important moments of my intellectual life.

There are countless scholarly debates on what exactly Hume himself thought, which I won't get into here. But what I can tell you with authority is that *I*—inspired by Hume—had come to accept the following two theses:

The Is-Ought Gap: You can't move in reasoning from facts that *aren't* about value or what you ought to do, to facts that *are* about value or what you ought to do.

Reason Ought to Be the Slave of the Passions: If you desire to pursue some goal, or feel it's valuable in your life, then, all things being equal, you ought to pursue it.

Once I had written them down next to each other, the tension was obvious. Suppose I strongly desire to become a professional philosopher. If reason ought to be the slave of the passions, then I can infer from the fact that I desire to be a professional philosopher that I ought to pursue the means of realizing this desire (assuming I don't have some stronger, conflicting desire). For example, I should work hard to get my PhD, publish in some reputable academic journals, etc. But if I make this inference, I am violating the Is-Ought Gap principle: I am moving from a cold-blooded fact about reality— that I happen to desire to be a philosopher—to a fact about what someone 'ought' to do—I ought to try to become a philosopher. The two principles at the core of my philosophical view of ethics were simply inconsistent.

There were basically three options open to me at this point. The most attractive option was to row back on my commitment to the Is-Ought Gap principle. I spent the next year or so in intense study of the various forms of 'value naturalism', the name for the broad family of theories which all hold that although value doesn't exist at the fundamental level of reality, value somehow emerges, perhaps from our desires, or from certain other observable facts about the natural world. It would require at least a book in itself to do justice to the many and varied theories that come under this banner, and I would encourage readers to explore this rich literature for themselves.[20] For my own part, I wound up dissatisfied with value naturalism, despite how much I wanted to believe it. The cold-blooded empirical 'is' facts about the natural world just seem radically different to evaluative facts about what we 'ought' to do, and so it's hard to see how there could be any kind of intelligible story about how the latter could emerge from the former. If you don't put value in, you can't get value out.[21]

Having rejected the comfortable middle-way option, at this point I found myself left with the choice between two extremes:

Value Fundamentalism: There are fundamental facts—as basic as the laws of physics or the axioms of mathematics—about what kinds of things are better and worse, and about how people ought to behave. If you can't get facts about value from other kinds of facts, then the value facts—if there are any—must be primitive facts in their own right. Perhaps, e.g., it is just a fundamental fact about reality that pleasure is good and pain is bad, or that sentient beings ought to be treated with respect.

Value Nihilism: There are no facts about better or worse, or about what ought to be done. In other words, value is an illusion.

When I was first confronted with this dilemma in my early 20s, value fundamentalism seemed like an incredibly extravagant position. We are entitled to believe in the findings of empirical science because they have been tested experimentally. But what on earth could entitle us to believe in a non-physical realm of value facts? Respect for Ockham's razor seemed to demand we shave off such a bold commitment.

And thus I became a value nihilist. For the rest of my 20s, I tried very hard to believe this position. Unfortunately, value nihilism is a hard act to pull off, as we'll find out in the next section.[22]

Can We Live without Value?

Let us be clear what value nihilism is. It's not the view that we can all make our own meaning and live meaningful lives on that basis. That's the subjectivist position I came to reject. Value nihilism is the view that every human activity is as pointless as counting blades of grass. We have a sense that certain things are worth doing: reducing suffering, advancing knowledge, creativity. But, if nihilism is true, our sense that these things are worth doing is an illusion. Everything is pointless.

The only appropriate response to nihilism is cynicism. Of course, nothing's going to stop you from continuing to explore and create, and to help other people. But if you're doing these from a sense that they're worthwhile things to do, you're a fool. If you decide to work for a development charity aimed at helping poor countries protect their tax base from predatory multinationals because you want to make the world a better place, you're a fool. There is no 'better' or 'worse.' There is just what is, and what is is utterly without value.

Actually, the first sentence of the last paragraph was not quite right. The word 'appropriate' is a value term, but the value nihilist doesn't believe in value. There can be no 'appropriate' or 'inappropriate' response to the world lacking value. Indeed, there cannot even be any 'reasons' counting in favour of doing one thing or another, as 'reason' is itself another value term, due to its logical connection to the word 'ought': what you 'ought' to do is determined by the balance of 'reasons'. It follows that for the nihilist, every act of deliberation is delusional, as deliberation is the act of weighing the 'reasons' for or against a certain course of action, and, according to nihilism, there are no such things as reasons.

Suppose I'm deciding whether or not to take a job in the US. I might think to myself: 'Well, the fact that I'll get more money counts in favour of going, but the fact that I'll spend time away from my family counts against.' I will then try to weigh those considerations, in order to work out what I ought to do. If nihilism is true, all of this is a totally confused activity. Talk of 'reasons', 'counting in favour', and what I 'ought' on balance to do, are riddled with value commitments. If there is no value, then deciding what to do by deliberation is equivalent to deciding what to do by reading tea leaves. Both are equally rooted in delusion.[23]

I said earlier that many philosophically inclined teenagers and philosophy undergraduates are often subjectivists. The braver ones may even identify as value nihilists. But in the latter case, more often than not, they are equating 'value' with 'moral value', and hence restricting their nihilism to principles concerning how we should or shouldn't behave in our treatment of others. They overlook the more subtle value claims that pervade our discourse, such as those involved in everyday, non-moral deliberation.

This kind of half-hearted nihilism, which only disbelieves in *moral* value, is perhaps liveable. After all, so long as the half-hearted nihilist happens to *want* to be kind to others, then she will probably judge that she has reason to be kind to others. But this slips back into the subjectivism discussed above. To fully embrace nihilism is to accept that you literally have no reason to do anything.

This philosophy cannot be lived out. Almost every human action involves appeal to reasons. Sometimes a person eats because they are literally compelled to do so by their animal hunger. But 99.99 per cent of the time, we eat something because we take ourselves to have reason to eat. The reason might be that we need to eat to stay healthy, or it might just be that it's going to be enjoyable. But if there are no reasons, if nothing counts in favour of action, then the only non-delusional human actions are the ones in which we are compelled by our animal urges. I'm currently enjoying my second child slowly transitioning from being a tiny animal that yields to every urge to a rational animal that can consider and deliberate what she wants to do. For the nihilist, this is essentially a process of being inculcated into a nonsense fantasy. I might as well be teaching my child about QAnon.

Nihilist philosopher and novelist Albert Camus beautifully captured the human situation according to nihilism by analogy with the Greek myth of Sisyphus, who was condemned by the gods to spend all of eternity engaged in the meaningless task of rolling a great boulder up a hill, only to see it roll down again.[24] If value nihilism is true, all human activity is just as pointless as this. However, Camus does not sanction despair but rather a heroic defiance, a brave determination to live optimistically in spite of the meaninglessness of the universe. This sounds admirable, even praiseworthy. Until one remembers that nothing is 'admirable' or 'praiseworthy' in a nihilistic universe. Camus is doing what so many half-hearted nihilists do: smuggling in value by the backdoor, in contradiction to their official worldview.

It gets worse. Value claims are pervasive not only in our discourse about action but also in our discourse about belief. We think people *ought* to apportion their beliefs to the evidence, that they *ought not* to believe contradictory statements, and so on. If there is no value, there is no reason to believe or disbelieve anything. The appearance of rational support for a given scientific theory is a delusion. In the light of this, we can see that it is impossible to take yourself to have reason to believe nihilism, simply because, if nihilism is true, there are no reasons. Value nihilism is truly the philosophy for the post-truth world. Nothing is rational or irrational. You can believe whatever you want.

Bart Streumer is a contemporary Dutch value nihilist who is far from half-hearted in his nihilism. Streumer has rigorously and analytically thought through the consequences of nihilism, working out what is or isn't consistent for a defender of nihilism to assert.[25] He argues, quite plausibly, that it is part of the nature of belief that one cannot self-consciously believe something whilst at the same time thinking there is no reason to believe it. But for the nihilist, there are no reasons to believe anything. It follows that, at least once you've reflected on the matter, nihilism is impossible to believe! Although in his writing he supports nihilism, Streumer refrains from saying he 'believes' the view he defends, confining himself to saying that 'the arguments point in that direction.'

Despite all of the above, I tried to live as a value nihilist for many years. It's impossible as a human being not to deliberate, not to weigh reasons for or against a given course of action. The trick to making nihilism bearable is to constantly remind yourself that even though all deliberation is a delusional activity, there is also nothing *wrong* with delusional activity, given that there's nothing right or wrong with anything. It's just something we do. A bit like humming.

Do Our Lives Have Meaning?

What's any of this got to do with cosmic purpose? What is the connection to the question we raised at the start of this chapter, namely: Can our lives have meaning in the absence of cosmic purpose?

Christian William Lane Craig and atheist David Benatar say no, as without cosmic purpose our lives lack cosmic significance. I suggested above that their concerns are exaggerated: that life is *more* meaningful if there is cosmic purpose, but it is not entirely lacking in meaning if there is no cosmic purpose.

The deeper concern is that value itself may be an illusion, and consequently that human meaning—which is a form of value—is also an illusion. To repeat: if nihilism is true, everything we do is as pointless as counting blades of grass. If you're a happy go lucky kind of person, you might think this is a terribly highfalutin thing to be worrying about, intellectual masturbation with no real bearing on everyday life. As long as you're happy with your lot, what's the problem? The problem is that when people really do feel that there is no meaning or significance in their lives, they are not happy. This is what depression feels like. If value nihilism is true, depression is the only true perspective on reality.

How does cosmic purpose help with this deeper concern? As we shall see in later chapters, the evidence we have for cosmic purpose is also evidence that value plays some kind of role in shaping the evolution of our universe. And if value plays a role in shaping the evolution of our universe, then value exists. In other words, while value could easily exist in the absence of cosmic purpose, the forms of cosmic purpose we find in our universe are unlikely to exist without value. Therefore, if we have evidence that there is cosmic purpose—as I argue in the next two chapters—we can be confident that value exists; an attitude of hope rather than despair is called for.

Thus, my philosophical journey has taken me from teenage subjectivism through youthful nihilism to my current position as a middle-aged believer in value. Moreover, despite my unthreatening appearance, I am a particularly rabid kind of believer in value. I still believe that the only plausible options are value fundamentalism and value nihilism. Given that the evidence for cosmic purpose casts doubt on value nihilism, we are left with value fundamentalism. Facts about better or worse, and about how we ought to live, are as fundamental as the laws of physics or the axioms of mathematics.

Many philosophers worry about how we could come to know such fundamental facts about value. We know about the empirical world through observation and experiments. By what faculty do we access the 'realm' of value?

This is a very serious problem, but it is also one that is strikingly similar to the challenge of explaining how we come to know truths about mathematics and logic.[26] My colleagues in the mathematics department discover facts about a timeless realm of numbers, functions, and sets. Paralleling the worry about our knowledge of value, we could equally ask by what faculty mathematicians explore mathematical reality.

My hope is that a unified story can be told about our knowledge of both the mathematical and the ethical structure of reality. But that is the story of another book. The conclusion for the moment is that the evidence for cosmic purpose can give us hope that value is real, and hence that the meaning our lives appear to have is not an illusion.

2

Why Science Points to Purpose

In the early days of the scientific revolution, most of its leading figures believed in God. Indeed, the conviction that a single divine mind was behind the workings of the universe underlay their belief in a law-governed order that could be discerned by the human mind. Newton even had God play a role in his cosmological theory, giving the planets a little nudge every now and again to maintain the stability of the solar system.[1]

However, as science progressed, there seemed to be less and less need for the God Hypothesis. Physics seemed to be revealing the universe to be a self-contained, self-sufficient system. A key figure in this development was the late 18th-/early 19th-century French physicist Pierre-Simon Laplace. Laplace managed to account for both the stability and the origins of the solar system without recourse to divine intervention. There is a well-known, although possibly apocryphal, anecdote of Napoleon questioning Laplace on where the Creator featured in his theory, to which Laplace allegedly responded, 'I have no need of that hypothesis.'

Even if God was not required to order the heavens, there still seemed a need for His guiding hand down here on Earth. In his famous watchmaker analogy, Laplace's contemporary William Paley argued that if we happened to find a watch lying on the floor, we would never dream of supposing that such a complex mechanism, perfectly suited to realize the function of marking the passage of time, had arisen by random chance processes, but would rather ascribe its creation to an intelligent designer.[2] By analogy, when we happen upon an organism, with its much more complex workings perfectly suited to realize the functions of movement, perception, digestion, and so on, we should likewise reject the idea that such a thing came about through the random interactions of particles. We should instead attribute its emergence to intelligent design.

Of course, even this role for God was removed by the arrival later in the 19th century of Charles Darwin's alternative explanation of the emergence of complex organisms. Darwin did not disagree with Paley on the absurdity of supposing that organisms were produced by random processes. Indeed, Darwin read Paley whilst studying at Cambridge, and later spoke very positively about his argument.[3] Rather, what Darwin had produced was an alternative both to God and to chance: natural selection. The basic idea is that, once there has emerged self-replicating life in a competitive environment, with occasional chance variations in characteristics, those organisms that happen to develop characteristics which yield a survival advantage will be more likely to survive than those that don't. Over large periods of time, this will result in the emergence of complex organisms moulded by the 'blind watchmaker'—to use Richard Dawkins' memorable play on Paley's analogy—of natural selection to be good at surviving.[4]

This was the final nail in God's coffin. There were now no longer any empirical grounds for believing in God, or anything like God. A short while later Nietzsche declared that 'God is dead.' Marx's view that religion was the opium of the masses, and Freud's view that the notion of God arose from a childish yearning for Daddy, gradually took hold among the intelligentsia.[5] The idea that science had displaced God, and eventually that science and religion are fundamentally opposed, gradually cemented itself in the public mind. For over a hundred years, there remained zero scientific evidence for God or cosmic purpose.

This all began to change in the 1970s. The painstaking work of putting together the 'standard model' of particle physics—our best theory of fundamental particles and the three forces of nature that dominate at the subatomic level—was finalized in the mid-1970s. The standard model, like most theories in fundamental physics, contains 'constants', i.e. fixed numbers we need to plug in to make the equations work. These include the masses of fundamental particles and the strengths of the forces governing them. Once we discovered these numbers, natural human curiosity raises the question of what the universe would have looked like if those numbers had been a bit different. As it happens, we are able to answer that question, by running computer simulations of universes with different numbers in their physics.

When scientists did this, what they found amounted to one of the most startling discoveries of modern science. The vast majority of the other universes generated through varying the values of the constants were incompatible with the existence of life. Not just carbon-based life of the form we are familiar with, but any kind of structural complexity whatsoever.[6] The strong nuclear force—the force that binds together the elements in the nucleus of the atom—can be represented by the number 0.007.[7] If that value had been 0.006 or less the universe would have contained nothing but hydrogen: the simplest element. If it had been 0.008 or higher, almost all of the hydrogen would have been burned off in the Big Bang and water would never have existed. In either case, there would be none of the chemical complexity we find in our universe.

The physical possibility of chemical complexity is also dependent on the masses of the basic building blocks of the matter that makes us up: electrons and quarks. For example, if the mass of a down quark had been greater by a factor of 3, again the universe would have contained only hydrogen.[8] As the astrophysicist Luke Barnes has remarked, chemistry exams in that universe would be easy to pass:

Question 1:   What's the only element? Answer: Hydrogen.

Question 2:   What's the only chemical reaction? Answer: Hydrogen atoms bond with hydrogen atoms to make a hydrogen molecule.

Contrast that with the more than sixty million known chemical compounds we find in our universe.[9] The situation would have been even worse if the mass of an electron had been greater by a factor of 2.5: the universe would have contained only neutrons, no atoms and no chemical reactions.[10] If we map out a space of possible universes generated by varying the masses of quarks and electrons, we find that only a tiny proportion of that space—the white space in Figure 1 on the following page—contains universes with interesting chemistry.

In fact, Figure 1 'zooms in' on a small region so that we can see the life-sustaining region. If we wanted to extend the axes up to the mass of the largest quark we've observed, we'd end up with a picture filling around 10,000 acres, almost all comprised of grey regions corresponding to lifeless universes.

This discovery has become known as the 'fine-tuning' of the laws of physics for life. It is important to note that this term doesn't imply a literal 'fine-tuner.' It is simply a way of referring to the fact that, for life to be possible, the values of these constants had to fall in certain very narrow ranges.[11]

The example of fine-tuning which has most shocked the scientific community arrived towards the end of the 20th century. It had been assumed since 1929 that the expansion of the universe was slowing down. But in 1998, physicists discovered that it was in fact speeding up, implying a hidden repulsive energy latent in space, which is now known as 'dark energy'. Moreover, when physicists measured the amount of dark energy in empty space—represented by a number known as the 'cosmological constant'—it turned out to be a little bit smaller than you'd expect, given the vast amount of matter that fills space. To be precise: a trillion, trillion, trillion, trillion, trillion, trillion, trillion, trillion, trillion, trillion times smaller! The precise value of the cosmological constant in the relevant units is:

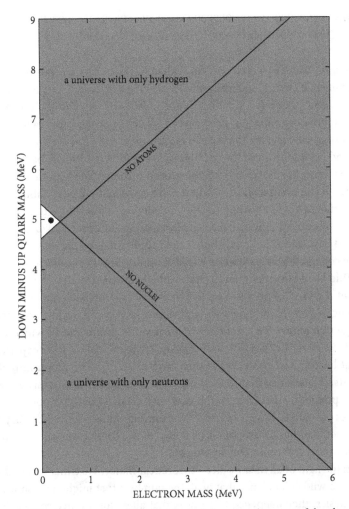

Figure 1 This image summarizes the effect of changing the mass of the electron and the difference in masses of the down and the up quarks. The gray region corresponds to universes with uninteresting chemistry; the tiny white region corresponds to universes with chemistry compatible with life. The black dot is our universe. We have 'zoomed in' on a tiny part of the possibility space, so that the life-permitting region is visible.

Based on figure 12 from Lewis and Barnes (2016), who based it on Craig J. Hogan (2009) 'Quarks, Electrons, and Atoms in Closely Related Universes,' in Bernard Carr (ed.) *Universe or Multiverse?*, Cambridge University Press.

0.00000000000000000000000000000000000000000000000000000000000000
000000000000000000000000000000000000000000000000000000000000000138.

Numbers like this—non-zero but ridiculously small—are not normal in fundamental physics, and many physicists find it very hard to rest content with the presence of such a number in our most basic story of the universe.[12] Nobel Prize-winning physicist Steven Weinberg called the value of the cosmological constant 'the bone in our throat.'[13]

And yet it's fortunate the cosmological constant is so small, because, if it hadn't been, the universe would have been incompatible with the existence of life. If the cosmological constant had been slightly bigger, things would have shot apart too quickly to allow gravity to clump things together into stars and planets. If it had been smaller than 0—in which case dark energy would have added to gravity rather than countering it—the universe would have collapsed back on itself in a split second.

This is all surprising and improbable, but what exactly does it tell us about the kind of universe we live in? What is so striking about the fine-tuning is that the numbers we find in our physics lie in the very narrow range that allows for a universe of great value, a universe in which there is life in all its richness, including people who can fall in love, experience great beauty, and contemplate their own existence. The fine-tuning does not itself guarantee that these things will emerge, but it is consistent with their possibility. In contrast, the vast, vast majority of other possible universes scientists have explored, by running simulations that vary the numbers in physics, contain little if any value. There is very little value in a universe containing only hydrogen.

In the light of all this, we face a theoretical choice. We either accept that it is just a wild coincidence that of all the numbers that might have shown up in physics, they happened to be ones that allow for a universe containing great value, or we hold that physics involves those numbers *because* they allow for a universe containing great value. I call the latter possibility the 'Value-Selection Hypothesis,' defined as follows:

Value-Selection Hypothesis: Certain of the fixed numbers in physics are as they are *because* they allow for a universe containing things of significant value.

The God Hypothesis would be one version of the Value-Selection Hypothesis, so long as God decided what numbers were going to show up in physics on the basis of trying to maximize the possibility of value in the universe. But if we can make sense of the idea of an impersonal force directed towards the good, selecting the optimal possibility from a range of options, then that could also be a version of the Value-Selection Hypothesis (we will explore such a hypothesis in great detail in Chapter 5). The essence of the Value-Selection Hypothesis is simply that, however the numbers associated with these constants were determined, it was done in order to allow for a universe containing things of value.

This then is our choice: incredible fluke or value playing a role in shaping our universe. What's wrong with the former option? Don't incredible things happen sometimes? There are widely reported happenings in which a form uncannily like the face of Jesus appears in toast:

Figure 2 Sometimes toast happens to pop out with a Jesus-shaped burn mark. Illustration: Emma Goff.

Should we avoid accepting that this was just a random improbable event by hypothesizing that Jesus is trying to communicate with us through toast? Of course not. As Carl Sagan famously said 'Extraordinary events require extraordinary evidence,' and the above image is nowhere near enough evidence for the extraordinary hypothesis that the son of God likes to chat through toast.

However, the reason we can accept the fluke option in this case is that an image of Jesus appearing in the toast by fluke is not so improbable. It's a bit improbable, which is why we enjoy seeing such images, but it's not *that* improbable. The trouble with the fine-tuning is that it is *really* improbable.

Because fine-tuning is rooted in physics that's abstracted from our everyday lives, it's not easy to get an intuitive feel for the probabilities involved. To make the kind of numbers we're talking about vivid, imagine you rolled a dice seventy times in a row and you got '6' every time.[14] While we're happy to accept that Jesus appearing in the toast is an amusing fluke, nobody who saw a dice coming up '6' seventy times would happily say, 'Ah well, it's just a fluke.' We'd think there was something fishy going on—perhaps a loaded dice—something that explains why the number 6 comes up each time. How does this compare with fine-tuning? The odds of getting '6' seventy times is 1 in 10^{55}; on a conservative estimate, the odds of getting a universe fine-tuned for life are 1 in 10^{136}.[15] Therefore, if it's irrational to put seventy optimal rolls of a dice down to chance, then it's certainly irrational to put the fine-tuning of our universe down to chance. In contrast to the 'Jesus Communicates through Toast' hypothesis, the rational pressure to accept the Value-Selection Hypothesis is overwhelming.

We can define cosmic purpose as any hypothesis according to which goal-directedness plays a fundamental role in determining what happens in the universe, in which at the fundamental level certain things happen *for the sake of some future goal*. The Value-Selection Hypothesis is a form of cosmic purpose, according to which our physics came to have the numbers it does—13.7 billion years ago—for the sake of the future goal of a universe containing things of great value. The fact that we have overwhelming evidence for the Value-Selection Hypothesis, therefore, entails that we have overwhelming evidence for cosmic purpose. Goal-directedness played some role in shaping the early universe.

To avoid misunderstanding, let me be clear that I am not suggesting we take the Value-Selection Hypothesis as a brute fact. It is natural to seek some deeper account of the process whereby some of the fixed numbers of our physics ended up being shaped by considerations of value; this is precisely what we will do in Chapters 5 and 6. Moreover, any adequate account here is going to need to explain *both* why the fixed numbers in question are appropriate to allow for positive value *and* why the universe also contains so much disvalue. These issues will not be neglected, but for now, I'll stick to defending the bare Value-Selection Hypothesis.

I imagine many readers will be suspicious at this point. If there really were scientific evidence for cosmic purpose, why is this not talked about more? Why doesn't the physics community overwhelmingly embrace the reality of cosmic purpose? On the contrary, we almost never hear scientists defending cosmic purpose. Whenever I raise the issue of fine-tuning with friends and family (I like to annoy people by engaging them in philosophical discussion at inopportune moments), in general, people have never even heard of it. This is probably because fine-tuning is not often discussed on popular science shows on TV and radio, such as *The Infinite Monkey Cage* in the UK or Neil deGrasse Tyson's show *Cosmos* in the US.

It is a curious situation. What I am discussing here is not controversial physics. The case I have made for cosmic purpose is not incontrovertible (I will explore the most discussed objections in the 'Digging Deeper' section). However, on the face of it, the data of contemporary physics seem to strongly support the Value-Selection Hypothesis, at least according to our standard mathematical definition of how evidence connects to theory (which we will also explore in the 'Digging Deeper' section). On this basis, the Value-Selection Hypothesis ought to be the default scientific view.

Why isn't it then? One factor is that the only version of Value-Selection Hypothesis that is much discussed is the traditional God Hypothesis, and I do think we have good reason to reject that hypothesis, as we'll discuss in Chapter 4. No doubt many, perhaps most, of those scientists and philosophers who consider cosmic purpose associate it with the God Hypothesis and reject it for that reason. However, as we shall explore in Chapters 5 and 6, there are other ways of accounting for cosmic purpose, ways which avoid the problems associated with the God Hypothesis.

More generally, there is strong cultural resistance not only to God but to any kind of purpose or goal-directedness at the fundamental level of reality. The enlightenment ideal is to dispassionately evaluate the evidence, and to apportion belief to whatever hypothesis is suggested by it. However, it is very difficult for human beings to do that. Over a hundred years with no evidence for anything like cosmic purpose has cemented in our culture a certain view as to what science is supposed to look like. People talk of religion as a crutch, and in many cases it is. But a certain kind of scientism, attached to a rigid view of what a scientific picture of reality 'should' look like, also gets into people's identity, into their sense of who they are. It's very reassuring to feel you're on the right team, on the side of reason unlike those other fools. And it's scary to have that worldview questioned if you're invested in it.

A large part of the reason cosmic purpose is not taken seriously in the scientific community is because it goes massively against the worldview that has come to be associated with science. To be clear, I'm not saying that the worldview was adopted without good reason. For a long time there was no evidence for cosmic purpose, so it was quite right that the scientific community rejected its existence, at least in so far as they were relying on scientific rather than philosophical considerations. But the evidential situation has now changed, and the culture should change with it. When the great economist John Maynard Keynes was challenged on the grounds that he'd changed his mind, he retorted 'When the facts change, I change my mind. What do you do, sir?' Quite right too. It's not easy though.

Scientists in the 16th century struggled to accept the mounting evidence that the earth was not, contrary to what had been assumed for thousands of years, in the centre of the universe. Popular science discussions often involve scoffing at this inability of our ancestors to follow the evidence where it leads. But every generation absorbs a worldview that it can't see beyond. In our own time, we are so used to the idea that science has done away with cosmic purpose that we are incapable of dispassionately considering the overwhelming evidence that has emerged in support of the Value-Selection Hypothesis. It may take time for the culture to catch up with the evidence. Future generations looking back will find it hard to understand how we ignored for so long what was staring us in the face: the clear and overwhelming evidence in support of cosmic purpose.

Choice Point: There are a few different options for where to go next. If you want to dig deeper into the science-based case for cosmic purpose we've been discussing in this chapter, perhaps because you have an objection you want answered, you can continue with the slightly more challenging 'Digging Deeper' section of this chapter. Alternately, you might want to go straight to the next chapter, which deepens the case for cosmic purpose through philosophical reflections on consciousness. Chapter 3 is a crucial part of the case for cosmic purpose, but it's also perhaps the most challenging bit of the book. If, for now, you have heard enough to be convinced of the case for cosmic purpose, you might like to go next to Chapter 4, in which I rule out the traditional Omni-God of Western religion as the source of cosmic purpose. Or you might like to go to Chapter 5, where I begin to explore other, in my view more plausible, options. I hope once you have a grip on the 'big picture' you may return to Chapter 3 to get the full story.

Digging Deeper

The Nature of Evidence

One of the courses I teach at Durham University is second year philosophy of religion. I suspect many students take it because they think it's going to be an easy ride, and that they can just rely upon the philosophy they learnt at high school. They are quickly disabused of that comforting thought when, in lecture two, they are introduced to the mathematics of probability. My approach is fairly unusual, but it seems to me unacceptable to be teaching students to discuss arguments in support of the existence of God without equipping them with the mathematical framework within which the entire debate is conducted these days. That framework is Bayes' theorem.

Thomas Bayes was an 18th-century statistician, philosopher, and Presbyterian minister. He came up with the theorem that took his name whilst wrestling with David Hume's argument against miracles. Hume had previously argued that, for any miraculous claim, it will always be more probable that the testimony resulted from error or deceit than from a genuine violation of the laws of nature. It was in thinking very hard about this argument that Bayes produced what has become a foundational equation of probability theory. It looks like this:

$$P\left(A\,|\,B\right) = \frac{P(A,B)}{P(B)}$$

Bayes' theorem is now crucial in many areas of science, from the predictive processing paradigm in contemporary neuroscience, to tracking the Covid19 pandemic. This is not the only time 'useless' philosophizing has had a highly useful spinoff. The philosophical theorizing of Gottlob Frege, Bertrand Russell, and Alfred North Whitehead at the end of the nineteenth and beginning of the 20th centuries resulted in the modern predicate logic that is so crucial for computer science.

Don't worry, we're not going to be looking at any more equations. But I do want to convey a basic understanding of how Bayes' theorem conceives of the relationship between evidence and theory. Bayes' theorem tells us that to find out whether the evidence we have supports a certain theory, we ask the following questions:

Question A: Supposing the theory is true, how likely is the evidence?

Question B: Supposing the theory is false, how likely is the evidence?

This is the bit that's most confusing to my students: that instead of asking how likely the *theory* is, we rather ask how likely the *evidence* is. But the point is very intuitive once you reflect on examples.

The Detective Example

Suppose you're a detective investigating a murder, and you want to know whether your theory that Joan is the murderer is supported by the evidence that Joan's DNA was found on the body. To find out whether this evidence supports your theory, you simply need to ask the two questions given above:

Question A: Supposing the theory is true (i.e. Joan is the murderer), how likely is the evidence (i.e. how likely is it that her DNA would be on the body)?

Question B: Supposing the theory is false (i.e. Joan is *not* the murderer), how likely is the evidence (i.e. how likely is it that her DNA would be on the body)?

If the answer to Question A is 'very likely' (or at least 'not unlikely')—because it's very easy to get your DNA on someone when you're strangling them—and the answer to Question B is 'very unlikely'—as Joan had no other interactions with the victim—then the evidence (that Joan's DNA was on the body) supports the theory (that Joan is the murderer).

We can sum this up with the general principle at work here (a principle which is derived from Bayes' theorem):

The Likelihood Principle: If the evidence is more likely assuming the theory is true than it is assuming the theory is false, then the evidence supports the theory.

Let's take another example for good measure.

The Rocks Example

Suppose you're walking along the beach, and you see rocks clearly arranged into the words 'LOVE LIFE LOVE THE MOMENT.' You form the theory that those rocks were deliberately arranged that way, as opposed to just being randomly washed up in that pattern. Again, to test this theory, we need to ask the two questions:

Question A: Supposing the theory is true (i.e. the rocks were deliberately arranged), how likely is the evidence (i.e. how likely is it that the rocks would be arranged in a meaningful sentence)?

Question B: Supposing the theory is false (i.e. the rocks were *not* deliberately arranged), how likely is the evidence (i.e. how likely is it that the rocks would be arranged in a meaningful sentence)?

It's far from inevitable that someone deliberately arranging rocks will arrange them in a meaningful sentence. They might instead make a smiley face, a pattern, or maybe they'll just arrange them in a random formation. But there's clearly a much better chance that a meaningful sentence will be formed if the rocks are arranged deliberately than if they were just washed up on the beach. The answer to Question A is 'not unlikely', and the answer to Question B is 'highly unlikely'. Thus, the evidence is significantly more likely assuming the theory is true (the rocks were deliberately arranged) than it is assuming the theory is false (the rocks were *not* deliberately arranged), and so the Likelihood Principle tells us that the evidence in this case supports the theory of deliberate arrangement.

Note that in applying the Likelihood Principle, we are always comparing *two* probabilities: the probability of the evidence assuming the theory is true and the probability of the evidence assuming the theory is false. It would get us nowhere just to point out how improbable it is that the rocks should arrange themselves into a meaningful word. After all, *any* arrangement of the rocks is highly improbable. Even if the rocks are just washed up randomly on the beach, it's still incredibly unlikely that, of all the possible locations of each rock, they should happen to wash up in precisely *that* random arrangement. The point is not merely that the rocks spelling out 'LOVE LIFE LOVE THE MOMENT' is improbable, it's that it's *more* improbable if the deliberate arrangement theory is false than it is if the deliberate arrangement theory is true. It was this way of comparing probabilities that was Thomas Bayes' genius insight.

The Likelihood Principle can be a bit confusing at first. But anybody can understand this if they want to, because all we're doing is noticing a principle we all implicitly appeal to in our thinking. If you saw rocks arranged to spell out a meaningful sentence, you'd immediately reach the conclusion that someone deliberately arranged them that way. And you'd reach that conclusion by implicitly applying the Likelihood Principle in your thinking. It's similar to how a native speaker of English implicitly grasps the rules of grammar when they speak, but finds it harder to grasp the rules when laid out in explicit form in a grammar lesson. It takes a bit of time to reflect on how the application of the rule matches what feels intuitively right to you when you reflect on examples. I recommend spending a bit of time reflecting on the two examples I have given above (the Detective Example and the Rocks Example) until you can see that the Likelihood Principle captures the intuitive conclusion we would all reach in these cases.

Once you've grasped the basic idea of the Likelihood Principle, we can apply it to the case of fine-tuning, to demonstrate the overwhelming evidence for what I above called 'the Value-Selection Hypothesis':

Value-Selection Hypothesis: Certain of the fixed numbers in physics are as they are *because* they allow for a universe containing things of significant value.

(In discussing the Value-Selection Hypothesis, there is a slightly awkward phrasing problem in that we need to talk about 'value' in two senses: (A) the 'value' of a constant, i.e. the specific number a given constant is measured to have, (B) 'value' in the sense of things being good. To resolve this ambiguity, I'm going to capitalize the first letter of 'Value' when talking about value in the second sense.)

To see that the fine-tuning supports the Value-Selection Hypothesis, we simply need to apply the Likelihood Principle. Recall that the Likelihood Principle tells us that to find out whether our evidence supports a certain theory, we need to ask the following questions:

Question A: Supposing the theory is true, how likely is the evidence?

Question B: Supposing the theory is false, how likely is the evidence?

Here the theory in question is the Value-Selection Hypothesis, and the evidence is the fact that the values of the constants are compatible with life. Let's answer these questions in turn:

Question A: Supposing the Value-Selection Hypothesis is true, how likely is that the value of the constants would be compatible with the existence of life?

Answer: **Not unlikely.** If the values of the constants are as they are in order to allow for a universe containing things of significant value, it's not so surprising that the constants have values that allow for life, given that life is of significant Value.

Question B: Supposing the Value-Selection Hypothesis is false, and the values of the constants were determined by processes not sensitive to considerations of Value (or perhaps it is just a fundamental fact about our universe that the numbers in physics are as they are), how likely is it that the values of the constants would be compatible with the existence of life?

Answer: **Incredibly unlikely.** Given that the vast majority of possible universes generated by varying the values of the constants are incompatible with the existence of life, it's incredibly unlikely that, by sheer chance, the constants in our physics would have values compatible with the existence of life.

In case you've forgotten it, the Likelihood Principle goes as follows:

The Likelihood Principle: If the evidence is more likely assuming the theory is true than it is assuming the theory is false, then the evidence supports the theory.

The above answer to Question A tells us that if the Value-Selection Hypothesis (the theory) is true, the fine-tuning (the evidence) is not unlikely, and the answer to Question B tells us that if the Value-Selection Hypothesis (the theory) is false, the fine-tuning data (the evidence) is incredibly unlikely. Hence, the evidence (i.e. fine-tuning) is more likely assuming the theory (i.e. the Value-Selection Hypothesis) is true than it is assuming the theory (the Value-Selection Hypothesis) is false, and thus— by the Likelihood Principle—the evidence of fine-tuning supports the Value-Selection Hypothesis.

The great thing about being able to appeal to Bayes' theorem is that we don't have to debate our 'intuitions' about the data, thereby allowing cultural biases to creep in. So long as the data is fairly probable if the Value-Selection Hypothesis is true and incredibly improbable if the Value-Selection Hypothesis is false, the Likelihood Principle will mechanically yield the result that the data support the Value-Selection Hypothesis.

The rest of the 'Digging Deeper' section consists of responses to objections.

Objection 1: The Multiverse Objection

I suggested above that there are significant cultural reasons why the scientific case for cosmic purpose is not taken more seriously. However, there is another factor we have not thus far discussed. Although many scientists and philosophers are impressed by the data of fine-tuning, and feel there must be some explanation of it, it is widely believed that there is a way of accounting for it that does not appeal to cosmic purpose: the multiverse hypothesis.

The attempt to explain fine-tuning in terms of a multiverse goes as follows. If there is only one universe, then it's incredibly improbable that, by sheer chance, it would have the right numbers for life. But if there is a huge, perhaps infinite, number of universes, each with slightly different numbers in their physics—in some gravity is weaker, in some electrons are heavier, and so on—then it becomes statistically highly likely that at least one will have the right numbers for life. If enough people buy a lottery ticket, it becomes very likely one will win. Hence, we have an explanation of there being a fine-tuned universe without need to appeal to cosmic purpose.[16]

The first thing to say about the multiverse explanation is that even if it provides *an* explanation of the fine-tuning, this is compatible with cosmic purpose *also* providing an explanation. If it turns out that there are two possible explanations of fine-tuning, then a proponent of one of these solutions is obliged to say why they opt for that solution over the other. Without some strong grounds to prefer one explanation over the other, we ought to remain somewhat agnostic over whether the fine-tuning we observe is the result of cosmic purpose or many universes. Perhaps the situation would be a little bit like the situation we find in quantum mechanics, where there are a number of hypotheses as to the reality underlying the data, with no clear way of deciding between them.

In fact, for a very long time I thought the multiverse hypothesis was on balance the superior explanation of the fine-tuning. Based on temperament alone, I would rather not be defending silly things like 'cosmic purpose.' However, I have taken a sacred vow to follow the evidence wherever it leads, and—over a long period of time—I was persuaded by conversations with probability theorists that the attempt to explain the fine-tuning in terms of a multiverse is rooted in confusion and flawed reasoning. The charge is that the multiverse explanation involves an error known as the 'inverse gambler's fallacy.' This has been discussed in the academic literature for thirty-five years,[17] but in a classic case of academics talking to themselves, almost nobody knows about it outside of academic philosophy. I was pleased a couple of years ago to finally get the point out to a broader audience via an article in *Scientific American*, and I hope to continue to spread the word in this book.[18] When it comes to the explanation of fine-tuning, there is no alternative to cosmic purpose.

Let's begin with the regular gambler's fallacy. You've been at the casino all night and have had a terrible run of luck. You think to yourself, 'I'll have one more roll. It's bound to be a double 6, as I've been playing for so long and it's really unlikely that I'd fail to roll a double 6 at least once in such a long session.' This is a fallacy, as the odds of your next roll being a double 6 are the same as the odds of any other individual roll being a double 6. How many times you've rolled in the same night is neither here nor there.

What about the *inverse* gambler's fallacy? Let's say you and I walk into a casino, and the first thing we see is somebody having an extraordinary run of luck, rolling double 6 after double 6. I say, 'Wow, the casino must be full tonight.' Confused, you ask me to clarify my line of reasoning. I say, 'Well, if there are only a few people playing in the casino tonight, it's improbable that you'd get somebody having such an incredible run of luck. But if there are very many people playing, it's not so improbable that one of them would happen to have a run of good luck.' This reasoning is also fallacious. We've only observed one person, and how many other people there are in the casino has no bearing on the odds of *the one person we've observed* rolling well. This is the inverse gambler's fallacy.

The multiverse theorist commits exactly the same fallacy. They are struck by the fine-tuning of our universe and conclude that there must be many other universes, most of which did not get the right numbers for fine-tuning. But all we've observed is *our universe*, and whether or not there are other universes has no bearing on the odds of *our universe* turning out to be fine-tuned.

Multiverse theorists standardly respond to this accusation by appealing to the 'anthropic principle', which is a fancy word for the truism that we couldn't have observed a universe with laws that weren't compatible with the existence of life (because if the laws weren't compatible with life, there'd be nobody around to observe them). This is of course true, but it is irrelevant. Suppose we add to the second casino story that a sniper is posted in the corner of the casino, ready to shoot me if the person playing by the door fails to roll incredibly well, just before I observe what the roll is. We now have a scenario in which I couldn't have observed anything other than an incredible roll (because if the player hadn't rolled well, I would've been shot dead before I observed it). But this makes not a jot of difference to my flawed reasoning. It's still a fallacy to infer from one person rolling well to a full casino. Likewise, it's a fallacy to infer many universes from one fine-tuned universe, regardless of the fact that we couldn't have observed a non-fine-tuned universe.

Moreover, we know why the inverse gambler's fallacy is a fallacy, namely, because it violates a very important principle in probabilistic reasoning known as the 'Requirement of Total Evidence'. This is the principle that when reasoning we should always work with the most specific evidence we have. To take an example from Paul Draper, suppose Jack is the defendant, and the lawyer for the prosecution shares with the jury that Jack carries a knife around with him.[19] In fact, as the lawyer for the prosecution well knows, Jack carries a butter knife around with him, but the lawyer chooses not to share this detail. Obviously the jury are going to be misled. It's not that they've been told a lie: it's true that Jack carries a knife around with him. The lawyer has misled the jury by giving them a less 'filled in' account of the evidence than is available.

The reason the inverse gambler's fallacy is a fallacy is that it also violates the Requirement of Total Evidence. The inference to the full casino starts from the evidence that 'someone in the casino has had an extraordinary run of luck.' But in the scenario described above, this is not the most specific evidence we have. The more specific evidence we have is that *this person* had an extraordinary run of luck. The Requirement of Total Evidence obliges us to take this more specific fact as our evidence, and once we do that, it no longer makes sense to infer to a full casino, because the number of people playing has no bearing on how likely it is that *this person* will roll well. Exactly the same problem applies to the multiverse inference. If we take as our evidence '*a universe* is fine-tuned,' then we are in violation of the Requirement of Total Evidence, as we have more specific evidence, namely that '*this universe* is fine-tuned.' And once we take this more specific fact as our evidence, it no longer makes sense to infer to a multiverse, because the existence of other universes has no bearing on how likely it is that *this universe* will turn out to be fine-tuned. The fact that we couldn't have observed a non-life-sustaining universe is neither here nor there; it's still not okay to violate the Requirement of Total Evidence.[20]

But isn't there scientific evidence for a multiverse? Yes and no.

The most popular scientific case for the multiverse begins with a hypothesis known as 'inflation', according to which our universe began with a brief period of rapid exponential expansion before slowing down.[21] Many prominent physicists think that inflation can explain some puzzling features of our universe, such as the fact that the microwave background radiation left over from the Big Bang is evenly distributed and fact that the space of our universe is pretty flat. Inflation in itself does not give us a multiverse: just one universe with a very rapid early period of expansion.

The multiverse comes in with the move from the basic thesis of inflation to the more specific thesis of *eternal inflation*. Although inflation has ended in *our universe*, some physicists hypothesize that our universe is just one small part of a vast mega-universe in which inflation never ends. According to eternal inflation, whilst inflation never ends in the mega-universe considered as a whole, various regions of it slow down to form 'bubble universes', our universe being one such bubble universe.

There is no direct evidence for eternal inflation. However, some physicists have argued that inflation is a bit like the magic porridge pot from the fairy-tale—once it starts, it's unlikely to stop! Whilst we don't really know what brought the inflation of our universe to an end—or indeed why it began in the first place—it is theorized that the end of inflation resulted from the vacuum (space considered independently of matter) undergoing a phase change, a bit like—very loose analogy!—the phase change when water boils and changes from liquid to gas. Given that inflation in our universe lasted a short period of time, one might think that the vacuum must have been disposed to change phase very quickly. Even so, given how rapidly space expands during the inflation phase, it seems that the rate of expansion will outstrip the rate at which the vacuum is disposed to change phase, with the result that there will always be some region of space—somewhere in the mega-universe—which is still inflating. If this is right, then inflation will go on forever, leading to the eternal inflation outlined above.

If the reasoning of the last paragraph is sound—which not all physicists accept—we now have a multiverse. But we don't yet have the right kind of multiverse to deal with the enigma of fine-tuning. Even accepting eternal inflation, there is no empirical evidence that the constants of physics—the strength of gravity, the mass of electrons, etc.—are different in these different bubble universes. And without such variation, the fine-tuning problem is even worse: we now have a whole multiverse that is fine-tuned for life!

At this point, many bring in string theory. String theory is even more speculative than inflation, and is taken seriously not because there is empirical support for it, but because many physicists think it is our best hope for unifying our best theory of very big things (general relativity) with our best theory of very small things (quantum mechanics). According to string theory, fundamental reality is made up not of particles or fields but of tiny one-dimensional strings, located in ten-dimensional space. The reason we see only three spatial dimensions is that the other seven are 'curled up.'

In the early days of string theory, it was hoped that there'd be only one way of curling up the extra dimensions, and that this would turn out to perfectly fit the physics of our universe. Unfortunately, it turned out that there was a huge number of possible 'compactifications'—ways of curling up the extra dimensions—corresponding to the physics of many different possible universes. A number commonly offered is 10^{500}. This range of possibilities is referred to as the 'string landscape.'

At first this lack of specificity in the theory was a huge disappointment. However, it later occurred to some physicists that this flexibility in string theory might help us in dealing with fine-tuning. Different options from the string landscape involve different values of the constants which, in our universe, are fine-tuned. Combining eternal inflation with string theory led to the speculation that random processes ensure that a wide variety of possibilities from the string landscape are realized in the different bubble universes. As the possibilities are gradually realized, there will eventually emerge a bubble universe with finely tuned constants.

Whilst there may be tentative empirical evidence for inflation, there is zero empirical evidence for different bubble universes exemplifying different options from the string landscape. As a philosopher, I'm not opposed to taking a certain hypothesis seriously on purely theoretical grounds. The trouble is that, as long as we respect the Requirement of Total Evidence, it follows that any view according to which fine-tuning arises by chance— including multiverse theories—is massively dis-confirmed by the evidence of fine-tuning.

To see this, let us define the 'cosmic fluke' hypothesis as follows:

Cosmic Fluke: Either the values of the constants in our universe were determined by chance processes or it is just a fundamental fact that the constants have the values they do.

The Cosmic Fluke Hypothesis is obviously inconsistent with the Value-Selection Hypothesis. According to Value-Selection Hypothesis, the constants have the values they do *because it is good* that they have those values; according to Cosmic Fluke Hypothesis, it is just chance that the constants have those values. The multiverse hypothesis outlined above is a version of the Cosmic Fluke Hypothesis: although it is not a fluke that *a universe* is fine-tuned—there are so many that it is highly likely that some universe will be fine-tuned—it is a fluke that *our universe* is fine-tuned. For any individual universe—including our own—it is incredibly improbable that that universe in particular would be fine-tuned.

So far we have appealed to the Likelihood Principle to assess the evidential support for a single theory. But it can also be used to compare two theories with respect to some evidence:

The Comparative Likelihood Principle: If the evidence is more likely assuming theory A is true than it is assuming theory B is true, then the evidence supports theory A over theory B.

The Requirement of Total Evidence obliges us to take the evidence of fine-tuning to consist of the data point that *this* universe—the only one we've ever observed—is fine-tuned. And that evidence is highly improbable if, as the Cosmic Fluke Hypothesis asserts, the value of the constants were determined by chance. The evidence that our universe is fine-tuned is massively more likely assuming the Value-Selection Hypothesis than it is assuming the Cosmic Fluke Hypothesis, and hence—by the Comparative Likelihood Principle—the evidence that our universe is fine-tuned massively supports the Value-Selection Hypothesis over the Cosmic Fluke Hypothesis. This does not rule out an eternal inflation multiverse, but it obliges us to adopt only versions of eternal inflation in which something is ensuring that a significant proportion of bubble universes are fine-tuned, for only on this supposition is it rendered likely that *this universe* is fine-tuned.

In summary: Even with all of the intricate and extensive resources of a multiverse, we cannot avoid the evidence for cosmic purpose.

Objection 2: The Most Common Objection Online

The most common objection one finds online to fine-tuning arguments is:

We shouldn't be at all surprised that the universe is fine-tuned for life. If it wasn't, we wouldn't be here to reflect on the matter!

This is often supported with Douglas Adams' puddle analogy (if I had a pound for every time this had been Tweeted at me I'd be very rich man...):

> imagine a puddle waking up one morning and thinking, 'This is an interesting world I find myself in—an interesting hole I find myself in—fits me

rather neatly, doesn't it? In fact it fits me staggeringly well, must have been made to have me in it!'[22]

We saw above how the multiverse theorist attempts to make use of anthropic reasoning. This objection is essentially anthropic reasoning on steroids, pressing that the anthropic principle—the obvious fact that we couldn't have found ourselves observing a universe that couldn't sustain life—somehow removes the whole problem. We can call it the 'Anthropic-To-The-Max' objection. The problem is that it's not clear *how* exactly this is supposed to remove the problem. Again, it's something that might seem to make sense superficially, but, once you actually focus on how evidence works—as expressed by the Likelihood Principle—it's baffling why anyone would think the anthropic principle makes a difference. Evidence is all about probabilistic connections between theory and evidence. The fine-tuning is strong evidence for the Value-Selection Hypothesis because:

If the Value-Selection Hypothesis is true, then fine-tuning is not unlikely.

If the Value-Selection Hypothesis is false, then fine-tuning is incredibly unlikely.

Whilst it is of course true that we wouldn't be here if the universe was not fine-tuned, that in no way undermines either of the above statements. And the two above statements are all that's needed for there to be strong evidence for the Value-Selection Hypothesis.

So the Anthropic-To-The-Max response makes no sense in terms of our theoretical understanding of probability. We can also cast doubt on it by analogy, the most famous of which is John Leslie's firing squad analogy:

The Firing Squad Analogy: You are to be executed by firing squad. Five expert shooters take aim at short range. They all miss. 'Wow, that was lucky,' you think, as they reload. The second time, again, they all miss. This continues, time and time again. You start to think something must be going on, as it's too much of a coincidence that all five would miss every time at close range. Perhaps you are being subject to some kind of cruel, mock execution, such as Dostoyevsky suffered. But then you remember: anthropic reasoning! If they hadn't missed every time, you wouldn't be around to reflect on it. Conclusion: you shouldn't be surprised at all, and certainly shouldn't try to explain why they keep missing.[23]

This seems entirely the wrong conclusion to reach, which, by analogy, has convinced many philosophers that the Anthropic-To-The-Max response to fine-tuning is misguided.[24]

There is perhaps another way, though, of reading Douglas Adams' puddle analogy, or what those who wield it on Twitter are getting at. Maybe the thought is that the fine-tuning is perfectly suited not for life in general, but just for the carbon-based life we are familiar with. Some examples of fine-tuning are indeed related to the possibility of carbon emerging. But carbon is special not just because we happen to be made up of it, but because it's a hugely versatile element. Vastly more compounds can be made by combining carbon with hydrogen, than can be made by combining hydrogen with other elements. In any case, the examples of fine-tuning we noted above are not to do with carbon but with the very possibility of chemical complexity or even of any complex structure at all. Whatever kind of life you're envisaging, it's not going to be able to exist if the universe collapses back on itself after a split second, which would have been the result of a negative cosmological constant.

It is sometimes pressed that we haven't explored *all* possible universes generated from varying the numbers in physics, and so can't know for sure whether further exploration of the possibility space would reveal life-permissible universes to be more common than we imagined. This objection reflects a very common strategy in responding to fine-tuning arguments, which is to ramp up the demands of proof to an artificially high standard that we wouldn't apply in any other case. Scientists have explored a lot of the possible universes generated from varying the values of the constants, and among those we've explored, the possible universes compatible with the existence of life are extremely rare. Of course, the evidence may change as we learn more, and perhaps we'll be surprised. But until that happens, we should work with the data we have.

Objection 3: Extraordinary Events Require Extraordinary Evidence!

Let's return to the case, discussed in the main section of this chapter, of an image of Jesus appearing in toast. Improbable things happen, and we shouldn't go around believing wacky things every time they do. Bayes' theorem allows for this, and to understand how, we need to bring in another aspect of Bayes' theorem: prior probability. The 'prior probability' of a hypothesis is defined as how likely the hypothesis is before we look at the latest evidence, just based on what we know about the world more generally. The overall probability of a hypothesis ends up being determined both by how strong the evidence is for it and by its prior probability. So a hypothesis with a reasonable prior probability—e.g. the hypothesis that I had cereal for breakfast this morning—won't need a great deal of evidence to end up having a high overall probability. However, a hypothesis with a very low prior probability—that Jesus communicates to people through their toast—is going to need a hell of a lot of evidence to get anywhere near likely. This is the more precise interpretation of Carl Sagan's famous phrase 'Extraordinary events require extraordinary evidence.'

Because of this constraining effect of prior probability, we don't need to worry that Bayes' theorem tells us that we get *some* support for wacky hypotheses all the time. Strictly speaking, the Jesus image in the toast is indeed more likely on the 'Jesus specifically wants to communicate with me through toast' hypothesis than on the 'A Jesus-shaped image appeared by chance' hypothesis, and hence, when we find an uncanny image of Jesus in our toast, this does supply *some* support for the former hypothesis over the latter. But if the prior probability of this hypothesis is somewhere around 0.0000000000001 per cent, then a little bit of evidence might get the probability up to 0.0000000000005 per cent, but it still won't end up being something we should take seriously. To put the point less rigorously but perhaps more intuitively: it's nice all things being equal to provide an explanation for events that would otherwise be pretty improbable, but if the explanation itself is even more improbable, then you're better off not believing it.

So can't we just react in the same way to fine-tuning? One might think:

> Sure, strictly speaking fine-tuning gives us *some* evidence for cosmic purpose, but cosmic purpose is so radically ill-fitting with our current scientific understanding of the universe that we should give it a miniscule prior probability. And so even if there is some evidence for it, it's still going to end up incredibly improbable.

The trouble with this response is that, in contrast to Jesus in the toast, the fine-tuning evidence for cosmic purpose is ludicrously strong. Bayes' theorem tells us that how strongly evidence supports a given theory is determined by how great the gap is between how likely the evidence is if the theory is true and how likely the evidence is if the theory is false. Let's consider the Jesus in the toast example in more detail:

<u>Question A</u>: Supposing the theory is true (Jesus likes to communicate through toast), how likely is the evidence (finding Jesus' image in toast)?

<u>Question B</u>: Supposing the theory is false (Jesus either doesn't exist as a supernatural being or doesn't have a preference for communicating through toast), how likely is the evidence (finding Jesus' image in the toast)?

Let's say for the sake of argument that the answer to Question A is 'highly likely'. If an omnipotent being wants to communicate through toast, then that's going to happen. What about Question B? Finding an image so uncannily like Jesus is a bit improbable. That's why we're entertained by this kind of thing. But it's not crazily improbable. The answer to Question B is 'fairly improbable'. So there is a gap between the two probabilities, and the evidence is more likely assuming the 'Jesus likes to communicate through toast' theory is true than it is if that theory if false. But the gap is not enormous, given that getting this kind of image by chance is not crazily improbable. And so the evidence we end up with, even if non-negligible, is not going to do much to boost the very low prior probability of the 'Jesus wants to communicate with me through toast' hypothesis.

In contrast, the probability of getting fine-tuned constants by chance is way-more-than-astronomically low. Physicist Lee Smolin estimates that the probability of getting fine-tuned constants by chance is 1 in 10^{229}.[25] This means that, even if you start off with a prior probability of 0.00000000000005 per cent for the Value-Selection Hypothesis, once you do the calculation, you're still going to end up with a probability for the Value-Selection Hypothesis of well over 99 per cent.

Objection 4: How Do We Know What the Probabilities Are?

Some of the deepest challenges to fine-tuning arguments question whether the probabilities they employ make sense, specifically the claims about the probability of the constants in physics having different values from the ones they actually have in our universe. Often, we work out how probable something is by investigating the world empirically. We know it is unlikely that a new-born baby will grow to be 8 feet tall, because we have observed many human beings and hardly any of them have grown to be that tall. But we've only ever observed one universe, and its constants have always had the same values, at least since the first split second of its existence. How, then, can we form a judgement about how probable it is that they would have had different values. Indeed, given that all our observations have conformed to physics with finely-tuned constants, maybe we should assign a probability of 100 per cent to the constants' being fine-tuned.

This objection implicitly assumes a *frequentist* conception of probability, which we can very roughly define as the view that how likely something is depends on how often it occurs. To say that there is a 50 per cent chance that heads will come up in a coin toss is to say that if we were to take all of the coin tosses in the world, we'd find that heads came up in half of them. On a frequentist view of probability, if you want to find out how likely something is, the thing to do is to go out and check how often it happens. Given that the values of the constants in our physics have never changed, this method doesn't allow us a way of ascribing a probability greater than 0 to the constants in physics having different values.

However, the plain fact of the matter is that it's impossible to do science with only recourse to a frequentist conception of probability, as it renders us unable to theorize regarding one-off events, such as continental drift, the evolution of humanity, or, indeed, the origins of the universe. One major source of evidence for continental drift was the similarity between animal and plant life on African and South America millions of years ago. Scientists made the plausible judgement that this kind of similarity would be very unlikely if the continental drift hypothesis were false but very likely if the continental drift hypothesis were true; by the Likelihood Principle, this fact of similarity supports the continental drift hypothesis. But how did scientists reach this probability judgement? On a frequentist account of probability, they would have to have observed many Earths with continental drift and many Earths without continental drift, and note that a greater proportion of those with continental drift had similar animal and plant life on Africa and South America. Of course, this is not something we are able to do, and hence scientists must have formed this probability judgement in some non-frequentist way.

How then do we reach probability judgements in the absence of statistical evidence about frequencies? An important part of the story is the *Principle of Indifference*, which tells us that, where we have a number of possibilities, with no reason to think one is more likely than the other, we ascribe an equal probability to each possibility. Suppose you're on the first episode of a gameshow, and the host tells you the prize is behind one of four doors, and you have to guess which. What are the odds the prize will be behind the fourth door? Given this is the first episode of the show, you can't work this out by watching all the old episodes and checking how many times the prize was behind the fourth door. What are you going to do? You will of course judge that there is a 25 per cent chance that the prize is behind the fourth door, and you'd be right to do so. This is an example of the Principle of Indifference in action. In the absence of any reason to ascribe a greater probability to the prize being behind one of the other doors (e.g. knowing that the host's favourite number is 2), you ought to ascribe an equal probability to each of the four doors hiding the prize.

It is exactly this same Principle of Indifference which is also being appealed to when we assign probabilities to the constants in our physics having certain values. Relative to a range of possible values of a given constant, in the absence of any reason to assign a greater probability to any particular possibility, we should assign an equal probability to each.[26]

But hold on, one might object, in the gameshow example, we don't know which door the prize is behind. That's why we're interested in working out how likely it is to be behind, say, the fourth door. However, in the case of the constants of physics, we *do* know what their values are, or at least we have strong evidence as to what their values are. How, then, can the same principle be being applied in the two cases? To answer this, we need to dig deeper into Bayesian probability theory, which we can do by responding to the next objection.

Addendum to Objection 4: At the time of reviewing the final proofs for this book, partly due to reading Paul Draper's manuscript 'Atheism and the Problem of Evil,' I now think about this a bit differently (although my view is not quite the same as Draper's). I would now argue that a hypothesis has an *intrinsic probability*, i.e. a probability independent of any evidence, that is dependent on considerations of simplicity and modesty. Hypotheses which are equally simple and modest have an equal intrinsic probability. And therefore, if there are, say, two equally simple hypotheses, neither of which has more supporting evidence than the other, we should assign to each an equal probability. Hence, we don't strictly speaking need to appeal to the Principle of Indifference, although the result is very similar.

Objection 5: How Do We Know the Constants Could Have Been Different?

The fine-tuning argument for cosmic purpose relies upon judgements about how likely it is that the constants in our physics would have the values they do. But how do we know the constants could have different values? Maybe there is some as yet undiscovered physical theory that entails that the constants have to have the values they do, in which case, perhaps there is nothing improbable about the way our universe has turned out.

This objection misunderstands the kinds of probability at play in Bayesian reasoning. They are not like the probabilities that we find in some interpretations of quantum mechanics, whereby past states of affairs don't guarantee what will happen in the future but rather yield *objective probabilities* of what will happen. There may be, for example, an 86 per cent chance that an electron will be found at a given location when it is observed.

These objective probabilities are not the kinds of thing we are interested in, at least not exclusively, when we are focusing on probabilistic connections between theory and evidence. In responding to the last objection, I raised the case of being on a gameshow, where the prize is behind one of four doors, and you have to guess which one. You would of course assign a probability of 25 per cent to the prize being behind the fourth door. But there is a definite fact of the matter as to which door the prize is behind, and hence what we are dealing with here is not an objective probability. The probabilities in question are *Bayesian probabilities*, and they concern what it is rational to believe.

We generally talk of belief as though it were binary: either you believe in something or you don't. But when we're engaged in Bayesian reasoning we're interested in *the degree to which one believes*. And when we're interested in belief as something that comes in degrees, we call it 'credence'. Often when you ask a philosopher whether they believe a certain philosophical view, for example, whether they believe in free will, they won't answer 'yes' or 'no', but rather will say something like, 'I have a 30 per cent credence in the existence of free will.' To say that there is a Bayesian probability of 25 per cent of the prize being behind Door 4 is to say that a rational person ought to have a credence of 25 per cent that the prize will be behind Door 4.

To dig a little deeper, when we apply the Likelihood Principle, we are interested not just in Bayesian probabilities, but in *conditional* Bayesian probabilities. When we're focused on conditional probabilities we're asking how likely something is hypothetically assuming something else is true. So, for example, we might ask how likely it is that the prize is behind Door 4, hypothetically assuming that Door 4 is the host's favourite number. Let's say the answer is 35 per cent. We'd express this by saying that the probability of the prize being behind Door 4 conditional on the host's favourite number being 4 is 35 per cent. In other words: if you knew that the host's favourite number is 4 (and had no other relevant information), you ought to have a 35 per cent credence that the prize is behind Door 4.

It feels more natural thinking about the above conditional probability if you don't know which door the prize is behind, and you're trying to work out the odds of it being behind Door 4 based on your knowledge that 4 is the host's favourite number. However, you could in principle still assess the probability of the prize being behind Door 4 conditional on the fact that 4 is the host's favourite number, even after the doors have been opened and you know which one has had the prize behind. To do this, you hypothetically ask yourself, 'Suppose I didn't know which door the prize was behind. How confident should I then be that the prize is behind Door 4, based on my knowledge that the host's favourite number is 4?'

Why on earth would you be interested in hypothetically ignoring some of your knowledge? Probably you wouldn't be in the gameshow case. But hypothetically ignoring some of your knowledge is often required when applying the Likelihood Principle. Here's a concrete example. One major source of evidence for preferring Einstein's theory of general relativity over Newton's theory of gravity is that it fits better with observable data concerning the precession of the perihelion of Mercury. The word 'perihelion' is the name for the point at which a planet comes closest to the sun. This changes over time, and that change is called the 'precession' of the planet's perihelion. The precession of Mercury's perihelion is measured to be 5,600 seconds of arc per century. This data was perfectly predicted by Einstein's theory, whereas Newton's was out by a shocking 43 seconds.

I just said that Einstein's theory 'predicts' this data, which might suggest that the data wasn't in when Einstein was formulated the theory of general relativity. But, in fact, this problem with Newton's theory had long been known about, and people had proposed various ad hoc ways of getting this data to fit with Newton's equations. Because we already knew that Mercury's perihelion precesses 5,600 seconds of arc per century, in order to apply the Likelihood Principle to judge which theory better fits the data, we need to hypothetically ignore this bit of data. It's only by doing this, that we can ask the following questions:

<u>Question A</u>:   Supposing Newton's theory is true, how likely is it that that Mercury's perihelion precesses 5,600 seconds of arc per century?

<u>Question B</u>:   Supposing Einstein's theory is true, how likely is it that that Mercury's perihelion precesses 5,600 seconds of arc per century?

The probability we get in answer to Question B is significantly higher than the probability we get in answer to Question A; thus, applying the (Comparative) Likelihood Principle yields the result that this data confirms Einstein's theory over Newton's.

We're doing exactly the same thing in the fine-tuning argument for cosmic purpose. To work out whether the fact that the values of the constants in our physics are life-permitting supports the Value-Selection Hypothesis, we need to answer the following two questions:

<u>Question A</u>:   Supposing the Value-Selection Hypothesis is true, how likely is it that the values of the constants are compatible with life?

<u>Question B</u>:   Supposing the Value-Selection Hypothesis is false, how likely is it that the values of the constants are compatible with life?

We already know that the values of the constants are life-permitting. And hence, in order to answer these questions—just as in the cases of the data concerning Mercury's orbit—we need to hypothetically ignore our knowledge of the data in question. We need to ask the following hypothetical questions:

<u>Question A</u>:   Suppose we knew that the Value-Selection Hypothesis were true, and we knew everything about the laws of physics and the initial conditions of the universe *except* for the values of the constants. How much credence ought we to have that the values of the constants are life-permitting?

<u>Question B</u>:   Suppose we knew that the Value-Selection Hypothesis were false, and we knew everything about the laws of physics and the initial conditions of the universe *except* for the values of the constants. How much credence ought we to have that the values of the constants are life-permitting?

Of course, the answer to Question A is that, in this hypothetical scenario, you should have a fairly high credence that the numbers will be life-permitting, and the answer to Question B is that, in this hypothetical scenario, you should have an extremely low credence that the numbers will be life-permitting. Hence, by applying the Likelihood Principle, we get the result that the fact that the values of the constants are indeed life-permitting strongly supports the Value-Selection Hypothesis.

One might worry that there's something a bit odd about imagining trying to assess how likely it is that the universe will turn out to be life-permitting. Given that anyone reasoning in such a hypothetical scenario would presumably be alive, we should surely take it to be 100 per cent likely that the universe is life-permitting! It's this kind of concern that tempts people into the kinds of anthropic reasoning discussed above. The concern is resolved once we appreciate that what we're really interested in are the objective relations of evidential support that hold between the evidence—*i.e. the fine-tuning of the constants*—and the theory—*i.e. the Value-Selection Hypothesis*; imagining a person assessing this probability is just a heuristic tool helping us do this. If it helps, we could imagine being a sort of disembodied spirit not reliant on the universe being fine-tuned for its existence, so long as we are aware that this is also just a tool for helping us to assess the relevant probability.

To return to the original objection of this section, once you understand what Bayesian probabilities are, the objection falls away. What we're interested in is not objective probabilities out there in the world, but how much credence one ought to have for certain statements, based on knowledge that certain things are true. If one day there arrives a grand unified theory which entails that the numbers in our physics that now seem arbitrary have to be the way they are, this could profoundly undermine the fine-tuning argument for cosmic purpose, although it will depend on whether or not there is fine-tuning in the most basic laws of this new theory. As previously noted, all we can ever do is work with the evidence we have at the present time, and the evidence we currently have strongly supports the Value-Selection Hypothesis.

Objection 6: Worries about Infinities

One other worry people sometimes have about the probability claims employed in fine-tuning arguments is that there seems to be an *infinite* range of possible values that any given constant could have.[27] If there are an infinite number of possible values that, say, the strong nuclear force could have had, and—in line with the Principle of Indifference—we assign an equal probability to each, then we reach the conclusion that the chance of the strong nuclear force having the value it does is one over infinity. This is an odd result: that something that happens to be true is infinitely improbable.

Not only that, if the range of possible values really is infinite, then the supposed empirical discovery that the range of values compatible with life is 'narrow' would cease to have any meaning. This is because, no matter how broad the range of values compatible with life is, compared to an infinite range of possible values, the probability of the strong nuclear force falling within the range compatible with life will be the same: one over infinity. Something seems to have gone wrong here.

However, it is not clear that the range of possible values of the constants is infinite, given that our current physical theories apply only below certain energy levels.[28] Moreover, it's perfectly reasonable to limit one's consideration to a certain finite range of constant values of the constants, so long as we don't 'cook the books' by considering a range which is arbitrary or contrived to give a certain result. Any Bayesian calculation involves taking certain assumptions as background information. To narrow our focus to a finite range of values for the constants, we simply build into the background information that the constants of our universe fall into that finite range. To object in this case to this perfectly standard aspect of Bayesian reasoning is again to make extra strong demands of fine-tuning arguments, demands that simply aren't made in other contexts.

3

Why Consciousness Points to Purpose

Warning: The philosophical subject matter of this chapter is inherently more challenging. On the other hand, it also contains some of the most original and important arguments of the book. I hope most readers will have a go at it, but if you want to skim/skip first time around, you can still get a rough understanding of the 'big idea' of the book based on the case for cosmic purpose in Chapter 2.

As a scientific community, we are not even at first base when it comes to explaining consciousness. It is now commonly accepted that accounting for how the brain produces consciousness is one of the deepest challenges of contemporary science. But still, many think we just need to plug away at our standard methods of investigating the brain and we'll one day crack it. What we are yet to fully absorb is that the challenge posed by consciousness is radically unlike any of the other tasks of science.

The reason is that consciousness is not publicly observable. I can't look inside your head and see your feelings and experiences. Consciousness is not something we discovered in a particle collider or looking down a microscope. We know that consciousness exists not from observation and experiment, but from our immediate awareness of our own feelings. If you're in pain, you're just directly aware of your pain. Moreover, the reality of one's own feelings and experiences is known with greater certainty than anything we know through experiments. Even though consciousness is not publicly observable, its reality is hard data that any adequate theory of reality must account for.

Of course, science is used to dealing with unobservables: fundamental particles, quantum wave functions, perhaps even other universes; none of these things can be directly observed. But there's a crucial difference when it comes to consciousness. In all other cases, we postulate unobservables *in order to explain what we can observe*. In the unique case of consciousness, *the thing we are trying to explain* is not publicly observable.

Please take a minute to meditate on my last sentence, and appreciate how it contradicts our contemporary understanding of how we find out about reality. It is standardly assumed that the only way to find out what the world is like is through public observation and experiment, by appeal to the empirical data we can all access through using our senses. If we can come up with some Grand Unified Theory that can account for all of the data of public observations and experiments, then the job of science will have been completed, and the resulting theory will be the best guess humans are capable of as to what reality is like. We can all give each other self-congratulatory pats on the back and get on with something else. But if you accept that consciousness exists, then there is something we know to be real that is not postulated in order to explain the data of public observation and experiment. And if it's real, then it needs to be accommodated in our overall theory of reality. In other words:

The reality of consciousness is a fundamental datum over and above the data of public observation and experiment.

Even if we can one day account for all of the data of public observation and experiment, the job of finding out about reality will not be complete. We will still be left with the task of accounting for the privately known reality of consciousness.

This is not to say that experimental science has no role to play in dealing with consciousness. Although I cannot observe your feelings and experiences, I can ask you how you're feeling. I can rely on your testimony about the unobservable world of your experiences. And if I do that while I'm scanning your brain, I can begin to map out how the observable facts of your brain are correlated with the unobservable reality of your consciousness. I can work out that such and such a pattern of neural firings goes along with such and such an experience. This is important data but it does not in itself explain consciousness. What we ultimately want from a theory of consciousness is an explanation of *why* brain activity goes along with experiences. Because consciousness is not publicly observable, this is not a question we can answer with an experiment.

Once it is appreciated that consciousness is not publicly observable, we are forced to choose between two radical options:

Deny Consciousness: If we want to preserve our current understanding of what science is, including its claim to be the unique way we find out about the nature of reality, then we must only believe in the things we know to exist through public observation and experiments. In this case, the privately known reality of conscious experience is an illusion: we think we have feelings and experiences, but we don't really.

Science Is Not the Full Story of Reality: There is more to reality that what we can learn about through experiments.

Of course, most people accept neither of these options. They want to eat their cake and have it too: accepting that consciousness exists, but without conceding the clear implication that there is something we need to account for that is not a postulation of public observation and experiment. This position makes no sense, and as a result the public understanding of consciousness is currently in a deep state of confusion.

The way out of this confusion is to make our choice as to which of the above options we go for. Some of my closest friends take the first path. Consciousness doesn't fit with our standard scientific approach, so we should dismiss its existence, just as we have dismissed the existence of witches and fairies.[1] I take the second approach: consciousness is real, and therefore we need to move to a more expansive conception of the project of finding out about reality. We could call it 'philosophy.'

How does 'philosophy' go about finding out about reality? How would we judge theories if not through the standard methods of experimental testing? In fact, people have an over-simplistic conception of how empirical enquiry works, as though it's some magical way of opening the hood of reality to directly apprehend the workings of the engine. In actual scientific practice, for any empirical data, there are always an infinite number of theories able to account for that data. Scientists choose the best theory from among this infinite set by weighing the 'theoretical virtues' of the various options, things like simplicity, elegance, and unity. The philosophical methodology I employ works in exactly the same way, just with an expanded data set. The task is to try to work out the simplest theory able to accommodate *both* the public data of observation and experiments *and* the privately known reality of consciousness. We might call this approach 'liberal naturalism': 'naturalism' due to its adopting of the familiar scientific method of trying to find the simplest theory that fits the data, 'liberal' in working with a more expansive conception of 'the data' that goes beyond just the data of public observation and experiments.

I am taking the time to spell this out because I am going to be making some bold claims in this chapter, and they are not going to be justified on the basis of public observation and experiments. I anticipate that the reaction of many readers will be one of skepticism. People generally think that you shouldn't believe bizarre theories of reality without hard data to support them. This is a wise attitude to have. But the theory I will propose here *is*, I believe, supported by hard data. It's just not the data of public observation and experiment. It is rather supported by privately known data concerning the reality of human consciousness. But this is also hard data: as certain as anything known through an experiment.

In my last book I defended panpsychism: the view that consciousness pervades the universe and is a fundamental feature of it. In this chapter, I will take this to the next stage by defending *pan-agentialism*, the view that the roots of agency are present at the fundamental level of physical reality. This is a hypothesis that has alienated even some of my panpsychist brethren, and I'm sure many will think this is the moment I jumped the shark. I would like not to be saying such strange things. But I've reluctantly reached the conclusion that the problem of consciousness is much deeper than anyone has thus far anticipated, and that adequately addressing it will push us far beyond our current picture of the universe.

The Problem of Meaning Zombies

Your consciousness is what it's like to be you. If you ask a scientist or a philosopher for examples, you'll standardly get a list like the following:

- Pain
- Seeing red
- The sensation of itchiness
- The taste of chocolate

These kinds of sensory experience are certainly the most vivid and easy to grasp examples of conscious experience. But there is more to human consciousness than these raw sensations. Human consciousness is also permeated with meaning and understanding. When I look out of the window in front of me—and I invite the reader to do the same—I don't just experience a meaningless mess of colours and shapes. I experience *people, cars, houses*. When I look at the screen of my laptop, I don't experience meaningless squiggles but *meaningful words*. If I look at my crying child, *her sadness* is evident on her face. In all of these examples, my understanding of what things are or mean is built into the character of my experience.[2]

Of course, the understanding of reality embodied in my consciousness is not always correct. I can hallucinate a pink elephant without there really being a pink elephant out there in front of me. My claim is not about how things are in reality but about how experience suggests things to be. My point is simply that an understanding of what things are or mean—that something is an elephant, that a certain word means 'cat', that someone is sad or excited—is part of the character of human experience. I will call this aspect of consciousness 'experiential understanding.'

We are so used to our experience being meaningful that we can easily overlook this aspect of consciousness, which is probably why it is little focused on in consciousness science. When we think about 'understanding' in this context, for example when thinking of artificial intelligence, we are more likely to focus on what we might call 'functional understanding,' a notion of understanding defined in terms of the behaviour of a system and its parts. In 1997, a computer program—Deep Blue—beat the chess grandmaster Garry Kasparov. If we're using the word 'understanding' to mean functional understanding, then Deep Blue has a deep understanding of chess. But if we're using the word 'understanding' to mean experiential understanding, then, assuming currently existing computers are not conscious, Deep Blue entirely lacks understanding of chess. When a human thinks about chess, they consciously understand the rules, possible moves, the rationale of a strategy. Deep Blue, in contrast, is just a mechanism that is blindly doing what it is programmed to do.

In 1950, the father of modern computing Alan Turing published a paper which laid out a way of determining whether a computer thinks.[3] He called it 'the Imitation Game'; we now call it 'the Turing test.' Turing imagined a human being engaged in conversation with two interlocutors hidden from view: one another human being, the other a computer. The game is to work out which is which. If a computer can fool 70 per cent of judges in a five-minute conversation into thinking it's the person, the computer passes the test. Would passing the Turing test—something which now seems imminent given rapid progress in AI—entail that the computer has thought or understanding? Turing's own answer was to dismiss this question as hopelessly vague, and to replace it with a new pragmatic definition of 'thought' whereby to think just is to pass the Turing test.

In this way, Turing defined into existence the idea of functional understanding.[4] The fact that this notion of understanding is now omnipresent in cognitive science shouldn't blind us to the fact that it is very different to the notion of understanding we standardly employ in ordinary thought, according to which understanding is a matter of what someone consciously grasps. Understanding a mathematical proof, or the punchline of a joke, or the solution to a puzzle, is the 'ah ha' moment when one 'sees' what the answer is. A non-conscious computer understands nothing in this sense. No matter how sophisticated its functional understanding, it totally lacks experiential understanding.

To make vivid the two very different notions of understanding at play here, I want to introduce the concept of a *meaning zombie*. The word 'zombie' is something of a technical term in consciousness research. We don't mean the lumbering, flesh-eating monsters we know from Hollywood movies. We're rather thinking of an imaginary creature which, in terms of its behaviour and the physical processing in its brain, is indiscernible from a normal human being but which totally lacks conscious experiences. If you stick a knife in a zombie, it screams and runs away but it doesn't actually feel pain. If a zombie crosses a road, it pauses and looks both ways, waiting for the traffic to clear, but it doesn't actually have any visual or auditory experience of the world around it. A philosophical zombie is just an unfeeling mechanism set up to behave like a normal human being. The Australian philosopher David Chalmers popularized the thought experiment of the 'philosophical zombie' as a way of exploring the conceptual possibility of bodies with minds.[5]

Meaning zombies are a development on this idea. In contrast to their regular zombie cousins, meaning zombies have conscious experience. But the conscious experience of a meaning zombie is restricted to meaningless sensation: colours, sounds, smells, tastes, etc. A meaning zombie has no experiential understanding of the world.

The 19th-century psychologist William James referred to the consciousness of an infant as a 'buzzing, blooming confusion.'[6] As the infant develops cognitively, its meaningless experience is gradually transformed into a world of people and things. In contrast, as a meaning zombie grows up, although it develops in terms of its behaviour and the information processing in its brain, eventually behaving just like a normal adult human being, its experience remains the blooming, buzzing confusion of the infant. When the meaning zombie looks around, it just experiences meaningless colours and shapes.

Of course, I am not suggesting there actually are any meaning zombies residing in the real world. It's just a useful way of illustrating that there are two very different ways in which we use the word 'understanding.' A meaning zombie has functional understanding, indeed it would pass the Turing test with ease. But it has no experiential understanding.

Meaning zombies are also useful for another reason. They pose a profound philosophical challenge: Why didn't we evolve as meaning zombies?

Natural selection is only interested in behaviour, as only behaviour matters for survival. A meaning zombie, by definition, would behave just like a real human being, and thus would survive just as well as a real human being. Natural selection has no interest in the quality of your inner life, so long as you're going to do the kinds of things that'll make you live longer and pass on your genes. On the face of it, therefore, we cannot explain in evolutionary terms why we are not meaning zombies.

This is a profound challenge. Perhaps more than anything it is experiential understanding that makes us human. It is our experiential grasp of what things are and mean, including the thoughts, feelings, and emotions of others, that connects us so deeply to reality. I would be reluctant even to call a meaning zombie—a creature whose entire experience of reality is a meaningless buzzing, blooming, confusion—a person. Certainly, I would say a meaning zombie has less moral significance than a rabbit, as the latter does have some experiential understanding of the world it engages with. It is experiential understanding that makes life worth living, and yet it's hard to see how our standard scientific account of how we came into existence—Darwinian natural selection—could explain the existence of experiential understanding.

The problem of meaning zombies is subtle and easy to miss. We want to say: Of course, if you have a meaningful understanding of the world around you, that's going to help you survive. This is surely true, but a meaning zombie's utterly meaningless conscious states, so long as they produce the same behavioural effects, will make it survive just as well. What is important for survival is *functional* understanding, not *experiential* understanding. I have no doubt that the behaviour that constitutes our functional understanding of the world results from the kind of consciousness we have, including our experiential understanding. But the behaviour that constitutes a meaning zombie's functional understanding would be rooted in the kind of totally meaningless conscious experience it has. Natural selection couldn't care less about the kind of experience that underlies behaviour, so long as the end result is the four 'F's: feeding, fighting, fleeing, and reproducing.

But does the idea of a meaning zombie make sense? Would, or even *could*, states of meaningless experience produce the kind of meaningful, intelligent behaviour exhibited by actual human beings? Wouldn't such states rather produce incoherent, senseless behaviour?

It depends. On the view I ultimately want to defend in this chapter, meaningful experience may be required for intelligent behaviour. But this cannot be taken for granted. On our standard scientific picture of things, what a human does is fully determined by the particles making it up, where the behaviour of those particles is fixed by the basic laws of physics. This is not to deny that the composite states of a human being, the states of its brain, for example, can impact the world. If my experiential understanding is an emergent feature of my brain, then it too may have a bearing on what my brain does. Still, what that emergent feature does will ultimately be traceable to facts about the particles making up my brain, acting in accordance with the basic laws of physics.

In other words, our standard scientific view involves the following commitment:

Micro-Reductionism: What a human being does is ultimately fixed by the fundamental particles making them up, and the behaviour of the fundamental particles making up a given human being is entirely determined by the basic laws of physics.

If micro-reductionism is true, a meaning zombie will behave just the same as an actual human being, so long as the ultimate story about the particles making it up is the same as that of an actual human being. The fact that meaningless, as opposed to meaningful, states of consciousness emerge at the higher level is neither here nor there.

As I've already said, meaning zombies are of course not real. In the actual world, wherever we find evolved organisms with high-level functional understanding, those organisms also have experiential understanding. But this just returns us to our question: what is it about our world that ensures that the evolution of functional understanding has gone hand in hand with the evolution of experiential understanding, given that these two notions of understanding are very different?

The first step to solving the meaning zombie problem is to deny micro-reductionism. If the basic laws of physics determine how I behave, regardless of whether or not experiential understanding emerges, then natural selection will not care whether or not experiential understanding emerges. It is only if the emergence of experiential understanding *makes a difference*, only if things with experiential understanding *survive better* than things whose behaviour is wholly determined by underlying chemistry and physics, that we can start to make sense of an evolutionary explanation of the emergence of experiential understanding.

But how could experiential understanding 'override' the laws of physics, in the sense of producing behaviour that is new and unexpected from the perspective of the basic laws of physics? Actually, there is a way of interpreting physics that can allow states of experiential understanding to bring about novel and unexpected behaviour without in any way 'violating' the basic laws of physics. We've had the problem; now let's turn to the solution.

The Revenge of Schrödinger's Cat

Because I'm a panpsychist who thinks consciousness is everywhere—more on this presently—people always think I'm going to like the interpretation of quantum mechanics whereby consciousness plays a role in 'collapsing the wave function.' For readers not familiar with this stuff, the oldie but still goodie tale of Schrödinger's cat is a good introduction. Schrödinger in fact came up with the story to show how ridiculous quantum mechanics (the theory he helped to develop) was. The idea is that we have a cat trapped in a box with a vial of poison hooked up to a tiny bit of radioactive substance. Due to quantum indeterminacy, it is equally probable that (A) some of the substance will have decayed in an hour—in which case the vial will be shattered and the poison released—and that (B) none of the substance will have decayed in an hour—in which case the vial will remain intact.

According to the interpretation of quantum mechanics that is taught in undergraduate physics courses, unobserved systems are very different to observed systems. While the box is closed and the system unobserved, the radioactive substance exists in an indeterminate state between having decayed and not decayed, and the cat exists in an indeterminate state between being alive and dead. When the box is opened and the system observed, the indeterminacy is somehow resolved: some radioactive substance either definitely decayed or definitely didn't decay, the cat is either definitely alive or definitely dead. This change from indeterminacy to determinacy is what is referred to as the 'collapse of the wave function.'

On the 'consciousness collapses the wave function' view, the unobserved system exists in this weird indeterminate state because there is no consciousness involved in it (let's pretend the cat was drugged into unconsciousness before the experiment was conducted).[7] It's only when an observer opens the box, and its contents become entangled with consciousness, that the indeterminacy 'collapses' into determinacy. This interpretation is no good for a panpsychist. Given that conscious particles are everywhere, including in the locked box with the unconscious cat, there would be no quantum indeterminacy anywhere. Schrödinger's cat would never get to be both alive and dead (or rather neither alive nor dead) at the same time.

Probably the most popular interpretation of quantum mechanics among philosophers of physics is the 'Many Worlds' interpretation, according to which what we think of as collapse of the wave function is actually the universe splitting into numerous branches.[8] In one branch, Schrödinger's cat is definitely alive; in another, the cat is definitely dead. By removing the collapse of the wave function, the Many Worlds interpretation removes the mysterious role for consciousness—which is one reason it's so popular—but the idea of a universe continuously branching into parallel worlds is pretty sexy in its own right. I don't like this version of quantum mechanics either, not least because it's hard to make sense of statements about probability when everything that could happen does happen.

My favourite interpretation of quantum mechanics is the most boring and unsexy of all of them: the *pilot wave interpretation*. The pilot wave interpretation of quantum mechanics gets rid of all the weirdness. If you're captivated by the idea that electrons are somehow both waves and particles, then you're not going to want to go for the pilot wave view, which squares this circle in the most prosaic way possible: there is *both* a wave-like thing *and* particle-like things. If you scoff at Einstein's declaration that 'God does not play dice' and delight in the idea of genuine randomness in nature, then you're going to be repulsed by the pilot wave view, which is utterly deterministic: the appearance of uncertainty is simply a reflection of our lack of knowledge as to the initial locations of particles. And if you're fascinated by the idea that the conscious observer plays a role in determining when the wave function collapses, then, you guessed it, you're heading for disappointment. As with the Many Worlds interpretation, the pilot wave view dispenses with collapse of the wave function.

We can think of the pilot wave view as what you get when you take the Many Worlds interpretation and add particles. What is this 'wave function' that either collapses or doesn't collapse? There are two ways of answering this question. On the one hand, what we call 'the wave function' is a bit of mathematical formalism, a function that yields complex numbers. On the other hand, on many interpretations of quantum mechanics, this formalism corresponds to a real constituent of physical reality, and that constituent of reality is also commonly referred to as 'the wave function'. In what follows, I will be adopting the latter understanding of 'wave function'; in other words, thinking of the wave function as something in reality itself rather than something internal to our theories of reality.

What is the 'wave function', understood as a constituent of physical reality? As with most things to do with quantum mechanics, scientists and philosophers disagree on how exactly it should be understood.[9] But on a fairly standard view, the wave function is an invisible physical field spread over a vastly high-dimensional space. How many dimensions? Take the number of particles in the universe, times by 3, and that's the number of dimensions of the space of the wave function. That's big, very big. Assuming it doesn't 'collapse', the wave function evolves deterministically over time in accordance with the core equation of quantum mechanics: the Schrödinger equation.

When you first hear about the Many Worlds interpretation of quantum mechanics, it sounds incredibly extravagant. However, Many Worlds theorists don't think of these universes as fundamental. Rather they typically think that the only fundamental thing is the wave function itself, and the three-dimensional world we experience—as well as the three-dimensional worlds experienced by all the other folk residing in 'worlds' that have branched off from our own—somehow emerges from the more fundamental reality of the wave function. (Find it difficult to make sense of how ordinary three-dimensional reality of tables, chairs, stars, and planets could emerge from the wave function? So do I, which is another reason I'm not a fan of the Many Worlds interpretation.)

The pilot wave theory accepts the existence of the wave function but also adds particles. This standardly involves positing two distinct spatial realms: the very high-dimensional space inhabited by the wave function, and the familiar three-dimensional space inhabited by particles. The wave function as a whole represents many different possible three-dimensional universes. In that sense, the space inhabited by the wave function can be thought of as the space of possibility. Given that Many Worlds theorists think that all that exists is the wave function, they have no way of distinguishing real from unreal possible universes; in fact there is no such distinction: they all exist! For the pilot wave theory, in contrast, of all the many possibilities represented by the wave function, only one corresponds to concrete arrangements of particles in three-dimensional space and thus is genuinely real.[10]

In addition to the Schrödinger equation shared by all versions of quantum mechanics, pilot wave theory postulates an additional equation—the guidance equation—which specifies a lawful connection between states of the wave function and locations of particles. On the face of it, the theory is telling us that the wave function pulls around the particles, a bit like—very loose analogy!—the way the moon pulls the tides.

There are challenges to the pilot wave interpretation of quantum mechanics, in particular other interpretations seem to fit more naturally with quantum field theory. However, pilot wave theory arguably has the philosophical advantage of being the least conceptually problematic of the various interpretations.[11] Whilst the solution to the meaning zombie problem I am about to describe fits well with the pilot wave theory, which is why I chose to frame it in those terms, it could be framed in terms of different physics if the pilot wave theory is ultimately rejected. My hunch is that we'll make progress on these foundational questions in physics when we get fully trained physicist-philosophers who take consciousness seriously.[12]

One worry philosophers, as opposed to scientists, have with the pilot wave interpretation is that it seems to leave particles causally impotent.[13] They are like dead weights with no causal power of their own, simply dragged around by the wave function. This is particularly worrying when you take into account that we are made up of particles, which can leave us with the idea that we are mere puppets whose strings are pulled by the wave function.

What I want to propose is a hypothesis which both addresses this internal difficulty with the pilot wave theory and solves the meaning zombie problem. I call it 'pan-agentialism.'[14]

Natural Agents

Here is the hypothesis I would like you to consider, as a way of solving the meaning zombie problem. As with my previous work, it's a form of panpsychism, according to which consciousness exists at the fundamental level of physical reality. For the sake of simplicity, for the moment I'll work with a particle-based interpretation of panpsychism, according to which the physical universe is made up of tiny fundamental particles, each of which has conscious experience of a very rudimentary form. Human experience is incredibly complex, but subjective experience comes in all shapes and sizes. If there is something that it's like to be a bedbug, then it's incredibly simple compared to what it's like to be a human being. There seems to be no inherent limit to how simple subjective experience could be. If particles have experience, then it is presumably of an incredibly simple form, corresponding to their incredibly simple physical structure. (In Chapter 6, we will explore a 'universe-first' rather than 'particle-first' conception of panpsychism.)

This new hypothesis—pan-agentialism—differs from the kind of panpsychism I've defended previously in holding that particles have a kind of proto-agency of their own. Particles are never compelled to do anything, but are rather disposed, from their own nature, to respond rationally to their experience.

Talk of a particle being 'rational' sounds ridiculous. This may partly result from the trend in Western philosophy at least since Aristotle to limit our concepts of 'rationality' and 'agency' to the kinds of things human beings do. Descartes took this to extremes: animals are mere mechanisms. Whilst in recent years, we have begun to extend the attribution of consciousness to non-human animals—fish, birds, reptiles—we have maintained the privileged status of being 'rational' for ourselves alone.

It is true that humans enjoy an astonishingly sophisticated form of rationality: the capacities for abstract thought, counterfactual reasoning, and logical deduction. But the first flowerings of reason are found in the most primitive drives. *To pursue what you're attracted to or to avoid what you dislike* are rational responses. It is rationally appropriate—all things being equal—to do what you feel like doing, and to avoid what repels you. To this extent, when simple organisms respond to their likes and dislikes, they are exhibiting a basic form of rational responsiveness, whether they recognize it or not.

Whilst most are now happy with the idea that many non-human animals behave in response to their conscious desires, we think of the inanimate world as functioning through a very different kind of causation. In the inanimate world, it is supposed, nothing yearns for anything. Things are simply compelled to act by other things: the red ball moved because the white ball hit it, the electron accelerated because the field was negatively charged. But it's not clear what justifies this assumption that what goes on in the inanimate world is so different to what goes on in us.

Empirical science tells *how* things behave, but it remains silent on *why* things behave as they do. We *assume* that inanimate entities are simply compelled to behave as they do. But it is consistent with observation to suppose that particles are engaging in a very rudimentary form of what organisms do: following their conscious inclinations. Of course, the conscious inclinations of an electron would be unimaginably simple compared to the conscious inclinations of even the simplest organisms. But there is nothing incoherent in the idea of conscious inclinations—the kind of thing we feel when we yearn for, say, food or water—existing in incredibly simple forms. And if particles do what they feel inclined to do, then they are thereby engaging in a very simple form of rational agency; or perhaps it might be better to say 'proto-agency.'

If particles are not compelled to behave in a particular way, wouldn't the inanimate world be a little less predictable? Not necessarily. I've spent the last six years raising young children. When an adult human yearns for something, they can rationally deliberate about whether it's the right thing to do, and may as a result choose not to act on their yearning. If a very young child yearns for something, in contrast, they're going to try and get it. Absent the capacity to rationally deliberate over whether eating the cookie is a good idea, it is for all intents and purposes inevitable that the child is going to go for the cookie.

To emphasize again, I don't think the child is acting irrationally. Doing what you feel like doing is one form of rational response, the simplest form. As a rational agent that lacks the cognitive capacity for more sophisticated forms of rational responsiveness, the child engages in the only rational response available to it: do what you feel like. Likewise, if a particle is disposed to respond rationally but lacks any kind of deliberative capacity, then it will inevitably respond rationally in the only way it is able: by following its inclinations. So long as the conscious inclinations of particles arise in a simple and predictable way, then the actions of particles will end up being simple and predictable.

But where do the conscious inclinations of particles come from? This is where the pilot wave theory comes in. My hypothesis is that the wave function does not drag the particles around, but rather causes particles to have certain conscious inclinations. The particles then freely respond with the most basic rational response: they do what they feel inclined to do. It is particles rationally responding to the conscious inclinations produced in them by the wave function that results in the standard predictions of quantum mechanics.

To flesh out pan-agentialism in more detail, we're going to need to add a little neuroscience into the mix.

Searching for the Neural Correlates of Consciousness

In the 1990s the neuroscientist Christof Koch bet the philosopher David Chalmers a case of fine wine that in twenty-five years scientists would have entirely pinned down the 'neural correlates of consciousness,' that is to say the forms of brain activity which are necessary and sufficient for conscious experience. Just as I was reviewing the final proofs of this book, Koch finally conceded defeat on this wager. He had little choice as there is zero consensus in this area.[15]

The lack of consensus is hardly surprising given that consciousness is not a publicly observable phenomenon, something which causes problems not just for addressing the 'why' question of explaining how consciousness comes to be in the first place but also the 'what' question of which physical states are correlated with consciousness. In order just to do the science it is necessary to establish 'detection procedures,' rules which specify how to bridge the gap from observable behaviour to the unobservable reality of consciousness.[16]

One superficially attractive candidate for a detection procedure is the following:

The Report Principle: If someone is experiencing something, they are able to report that they are having that experience.

Figure 3 Compare this with Figure 4 without looking at the two images side by side. There is a difference between the two pictures but most people are unable to tell what it is. Does it follow that we do not experience the part of the image that changes? Scientists and philosophers disagree on this, which leads them to very different theories of consciousness.

Thanks to Ned Block for suggesting this example from Block 2008, and for Ron Rensink for allowing me to use this image.

Figure 4

With the Report Principle in place, we can ascertain whether someone is experiencing something by simply asking them. And if we do this in tandem with observing which bits of their brain light up when they have the experience they claim to be having, we can in principle establish which kinds of brain activity are correlated with the existence of the experience in question. This is a gross over-simplification but it serves to illustrate in broad brush strokes how the Report Principle can facilitate progress in pinning down the neural correlates of consciousness.

The trouble is that the Report Principle is controversial. One reason for this is the phenomenon of change blindness. Intuitively, when we look at a picture, we feel we are visually experiencing all of the detail of the picture. However, there is overwhelming evidence that subjects can fail to notice significant changes in an image they are seeing at close range, even when those changes take up a sizable part of the image. There seem to be limits on how much detail we are able to attend to in what we intuitively take ourselves to be seeing. Let's take an example. Before reading on, can you tell the difference between the image on p. 61 and the image on p. 62, without looking at them side by side?

Change blindness raises an interesting question: Am I actually experiencing those aspects of an image which change without my noticing? In case you couldn't work it out—it took me a while!—the first image contains a big propeller under the wing nearest the viewer, which is missing from the second image. When shown the images consecutively (with a blank image briefly shown in between), subjects are typically unable to tell that there is a difference, even after being shown the images many times. Does it follow that the subjects do not in fact experience the propeller when viewing the first image? After all, it's odd to think someone could be conscious of the propeller, then be conscious of the absence of the propeller a moment later, without noticing that it's disappeared. Some science-osophers (a term I've just made up as it's not clear whether researchers engaged in these debates are doing science or philosophy) do indeed want to say that if I'm unable to access information about something I'm seeing—in this case, whether or not there's a propeller— then that information doesn't really feature in my conscious experience. In other words, I don't really experience the propeller. In fact, science-osophers on this side of the debate tend to hold that we only experience about four things at a time, despite our sense of having a much richer visual experience.

Other science-osophers see no problem with the idea of there being aspects of my experience that I can't access information about. Consider, for example, cases where one is driving on autopilot with one's mind on other things. Some science-osophers claims that we do experience the road— that's why we don't crash!—but without being reflectively aware of what we're experiencing or able to report on it. Indeed, one will often later be entirely unable to say what happened on the journey. The term 'Overflow' has been introduced to refer to the thesis that our conscious experience is richer than what we are able to report on. Science-osophers who accept Overflow deny the Report Principle (as, by definition, overflowing con-sciousness can't be reported on) and conversely those who accept the Report Principle deny Overflow.

These two opposing teams of science-osophers end up having wildly dif-ferent views on where consciousness is located in the brain. Those on the side that denies Overflow and accepts the Report Principle generally assert a tight connection between consciousness and *reflective cognition* (roughly, the brain's ability to access and use information to make decisions), and thus believe that the prefrontal cortex—where high-level cognition takes place—is essentially involved in all conscious experience. Those on the side that accepts Overflow and denies the Report Principle, in contrast, tend to locate sensory conscious experiences in whichever local bit of the brain the relevant sensory processing takes place. This dispute is commonly characterised as a disagreement as to whether consciousness is located at the front or the back of the brain.[17]

I think there is a way forward in this debate. We have focused a lot on the negative point that consciousness is not publicly observable. But the flip side of this is the positive point that we know about consciousness in a very direct way, through our immediate awareness of our own feelings and experiences. I believe this way of knowing about consciousness 'from the inside' yields knowledge of its nature than can be helpful in making progress on the hunt for the neural correlates of consciousness.

One thing I think we know about consciousness 'from the inside' is that it cannot be vague whether or not a system is conscious. The word 'vague' is a semi-technical term in philosophy, which refers to phenomena that admit of fuzzy, borderline cases. Baldness is a vague phenomenon. There is no precise number of hairs that marks the boundary between people who are and people who are not bald. Someone with a full head of hair is definitely not bald. Someone who has no hair at all is definitely bald. But there are also people who have had significant hair loss so that we couldn't say they are definitely in the 'not bald' category, whilst at the same time they have not had quite enough hair loss to say they're definitely in the 'bald' category. I myself, for a little while longer at least, am in this fuzzy, borderline grouping.

Most of the things we talk about are vague: being tall, being old, being rich, being happy, and so on. There is no exact minute when someone transitions from being young to being old, no exact penny that makes one transition to being rich, etc. But I don't think we can make coherent sense of the idea of consciousness being vague. To take a concrete example, borrowed from Eric Schwitzgebel, consider the case of the humble garden snail.[18] It is a matter of some controversy whether or not a garden snail has consciousness, whether or not there is something that it is like to be a snail. Whilst our knowledge of whether or not the snail is conscious is highly uncertain, there surely must be a fact of the matter as to whether or not the snail is conscious. Maybe the snail has a very simple form of subjective experience. Maybe it doesn't have any subjective experience at all. What doesn't make sense is to hold that there's no fact of the matter as to whether or not the snail has subjective experience, in something like the way there's no fact of the matter as to whether or not I am bald. Either there is something that it's like to be a snail or there's nothing it's like to be a snail. The very concept of subjective experience rules out borderline cases of consciousness.[19]

This is very useful information, drawn not from experimental science but from what we know about consciousness 'from the inside,' in virtue of being conscious. Although it is hard to decide on experimental grounds which is the correct theory of the neural correlates of consciousness, this non-experimental data point that there cannot be borderline cases of consciousness provides a way of whittling down the options.

Consider, for example, one leading contender: the global workspace theory. According to this view, information in the brain becomes conscious when it is available not only *locally*—to one specific system in the brain—but *globally*—to many different systems in the brain, including perceptual systems, long-term memory, and motor control.[20] On this view, to steal a metaphor from Daniel Dennett, information is conscious if it's 'famous' in the brain: if lots of different systems know about it.[21] But how famous does information have to be in order to be conscious? To how many cognitive systems does it need to be available? There is surely not going to be some exact threshold at which information becomes famous enough to become conscious. And hence there are going to be some organisms—perhaps snails—whose central nervous systems embody information that is in the borderline region between being famous enough to make the system conscious and failing to be famous enough to make the system conscious. If the global workspace theory is true, there will be no fact of the matter as to whether such systems are conscious. Conversely, if—as I suggested above—it makes no sense to suppose that there could be no fact of the matter as to whether or not something is conscious, it follows that the global workspace theory doesn't make sense either.[22]

In fact, as far as I know, there is only one extant theory of the neural correlates of consciousness that passes this test: the integrated information theory.[23] This theory revolves around a certain measurable characteristic of a system, which the theory refers to as 'integrated information'. We don't need to worry too much about what integrated information is, but we can roughly think of it as how much the system's parts work together to constrain its possible past and future states. For example, the retina of the eye has low integrated information, as its state at any given time is compatible with a wide variety of possible states at the next moment, dependent on what light hits it. The brain, in contrast, has very high levels of integrated information, as the range of states it could enter into at the next moment is very narrow.

The crucial point for our purposes is that this theory offers us an utterly precise threshold that marks the difference between a system having and lacking consciousness. According to integrated information theory, at the precise moment when there comes to be more integrated information in the system as a whole than there is in its parts, the system becomes conscious. Conversely, at the exact moment when there ceases to be more integrated information in the system as a whole that there is in its parts, the system ceases to be conscious. By avoiding the incoherent idea that some organisms are neither definitely conscious nor definitely non-conscious, integrated information theory may be the only coherent proposal currently on offer concerning the neural correlates of consciousness.

What has been so exciting for me in recent years is to see scientists and philosophers coming together to lay the foundations for a new approach to consciousness. Giulio Tononi is the neuroscientist who came up with the integrated information theory of consciousness. Hedda Hassel Mørch is a philosopher who spent a year in Tononi's lab, developing a panpsychist version of the integrated information theory, according to which at the moment conscious entities combine to form a system in which there is more integrated information in the system as a whole than in any of its parts, those conscious entities fuse together to form a unified conscious system, ceasing to exist as separate conscious minds in the process.[24] The Mørch-Tononi view is exactly the kind of thing we need to make progress on consciousness, a unified theory that answers both the 'what' and the 'why' questions of consciousness. I have my own version of panpsychism, which we'll get to in Chapter 6, but for the moment we'll work with the Mørch-Tononi view.

Freedom Evolves

I want now to bring everything together, by combining the Mørch-Tononi hypothesis described in the last section with the pan-agentialist theory developed in the section before that.

According to pan-agentialism, matter is inherently disposed to respond rationally to the reality that is presented to it. Particles simply follow their conscious inclinations. This is rational behaviour—it's rational to do what you feel like doing—but of an incredibly simple form. But as complex conscious minds emerge, they start to have experiential understanding of the world around them: at first, the most basic forms of cognition. If particles have conscious inclinations, then they are directed at very simple, immediate goals: I want to do *this* and I can do *this*. But, as life evolves, the conscious inclinations of complex minds start to become embedded in a complex web of meaning and understanding. The tiger doesn't just desire *do this now*, it desires food, drink, and sex. Whilst it doesn't have the rich conceptual understanding of these biological realities that a human being has, its representation of them is still incredibly complex. These objects of desire are not immediately present but rather occupy locations in a complex representation of the external world across space and time.

My proposal is that *the very same capacity for rational responsiveness* possessed by the tiger is also possessed by the particle. But whereas the particle can do very little with that capacity, when that capacity is married to the rich cognitive understanding of the tiger, it flowers into a complex engagement with the world around it. The potential for agency was always latent in the proto-agency of the particles, but it needed to combine with experiential understanding to come to fruition:

proto-agency + experiential understanding = agency

What of human beings? Certainly our experiential understanding of the world is much greater than non-human animals, resulting in a richer form of agency. Some philosophers think that our agency is just a more sophisticated form of what the tiger has, to the extent that—as with the tiger—human agency is ultimately simply a matter of pursuing our conscious inclinations: doing what we feel like doing. We are back to David Hume's view that 'Reason is...the slave of the passions', but this time understood just as a claim about our psychology rather than about how things ought to be.

I'm inclined to think there's more to human agency than this. At least part of the time, we do things because we recognize that they are *worth doing*. This might reflect some grand purpose: advancing human knowledge, or helping find a vaccine for Covid19. Or it may simply be choosing the thing on the menu that will bring me most pleasure, realizing that it's good to feel pleasure. Such cases, I think, refute Hume. It is not *desire* that is determining action but *judgement*, specifically the judgement that a certain action is objectively worth doing.

Whether you agree with me or Hume on this may make a difference to your view of free will. If our actions are determined by our conscious inclinations, with the strongest inclination inevitably winning out, we may end up with a purely deterministic story of human action. But if humans have the capacity to choose whether to follow their desires or their value judgements, if a human can pursue something not because they desire it but because it is *worth doing*, then there may be nothing that determines, prior to the moment of choice, which action is decided upon. This kind of radically undetermined free choice is what philosophers call 'libertarian free will.'

In the 'Digging Deeper' section of this chapter, I will argue for the coherence of libertarian free will.

The pan-agentialist view defended in this chapter is not necessarily committed to libertarian free will. A pan-agentialist may hold that all matter is disposed simply to follow its conscious inclinations, with the strongest inclination inevitably winning out, resulting in a kind of determinism. However, if there are objective facts about value, which the kind of cosmic purpose defended in this book commits us to, then presumably human beings are able to recognize and respond to considerations of value: I can do something because I think it's *worth doing*, regardless of whether or not I *feel like* doing it. For this reason, it seems to me more natural for a pan-agentialist to hold that recognition of value brings with it libertarian free will, the capacity to choose whether to respond to one's inclinations or to one's judgements about what is worth doing. For example, I might feel a strong desire to lie around watching TV but at the same time judge that my time would be better spent reading some philosophy. Free will consists in my capacity to choose whether to follow my desire or my judgement.

For this kind of pan-agentialist, there are three stages in the evolution of agency:

1. Proto-agency: The capacity of a particle to respond to its immediate inclination to perform a specific action in the here and now.
2. Agency (proto-agency + experiential understanding): The capacity of an organism to pursue objects of desire spread out over space and time.
3. Free will (proto-agency + experiential understanding + recognition of value): The capacity of a human being to choose whether to respond to their conscious inclinations or to their value judgements.

Note that in each case, there is not some magical new capacity that appears totally out of thin air. Rather, the basic capacity to respond rationally—that very capacity possessed by the humble particle—has latent within it all of these possibilities. But these possibilities can be realized only when conjoined with the right kind of conscious understanding.

Freedom from Physics

Some of the above may sound like truisms. Of course the tiger pursues its desired goals of eating, mating, etc. Who would say otherwise? What is crucial for the pan-agentialist view is that the act of responding to one's value judgements or conscious inclinations—whether one is a particle, a tiger, or a human being—is a *fundamental form of causation*. On the contrasting view of micro-reductionism, everything that happens in the biological world is ultimately determined by what's going on at the fundamental level of reality. Although we can truly describe the tiger as 'chasing the gazelle because it's hungry and understands that this is a way to get food,' this is a just a higher level description of processes that can in principle be exhaustively described at the level of particles and fields.[25] The point is perhaps clearer with an analogy. When I say, 'The party last night was wild!,' this is just another way of describing what certain people were doing; there's no such thing as the 'actions of the party' over and above the actions of the partiers. Similarly, for the micro-reductionist, talk of tigers chasing gazelles is just another—much more convenient—way of describing complex interactions of particles.

In contrast, according to pan-agentialism, as complex conscious systems with experiential understanding begin to emerge, they bring into being new causal principles over and above the basic laws of physics. Physical entities are no longer just responding to the very basic inclinations imparted to them by the wave function. They are rather responding to their conscious understanding of reality, and their attractions to certain goals within that reality. The result will be systems that behave in ways that depart from the predictions associated with quantum mechanics, which are generated by a principle known as the Born rule.[26] But on the hypothesis currently under consideration, this should not be seen as something which is inconsistent with quantum mechanics, at least not the interpretation of quantum mechanics I have outlined here. For what the Born rule tracks, on this view, is how physical things behave when they're responding to the very simple inclinations imparted to them by the wave function. When there emerge physical systems which can grasp more sophisticated reasons, e.g. their future well-being rather than immediate gratification, they may act in other ways. But this is what we'd *expect*, given the theory—not just the bare equations but the interpretation of what is going on in the underlying reality.[27]

But don't we know empirically that the Born rule—the principle which generates the predictions standardly associated with quantum mechanics—accurately captures what goes on the brain? The truth is we don't know anywhere near enough about what goes on in the brain to know whether or not this is true. We know a great deal about the basic chemistry of the brain—how neurons fire, how chemical signals are transmitted, etc. And we know a fair bit about the large-scale functions of the brain, i.e. what different bits of the brain contribute with respect to the overall functioning of the body. What we are almost clueless on is how large-scale functions are realized at the cellular level. In other words, we are almost totally ignorant about *how the brain works*. We'd have to know a lot more about how the functions of the brain are realized before we could be confident that everything that happens in the brain is reducible to underlying chemistry and physics.[28]

Moreover, there is empirical work pushing in an anti-reductionist direction, arguing for cases of 'strong emergence,' i.e. causal dynamics that are not—even in principle—explicable in terms of fundamental physics. Perhaps the best cases for strong emergence arise in *condensed matter physics*, the study of how matter behaves in low-energy environments where matter 'condenses' into atoms and molecules. In 1998 Robert Laughlin was awarded the Physics Nobel Prize (along with Horst Störmer and Daniel Tsui) for the discovery of the *fractional quantum Hall effect*, a surprising principle of organization we find among electrons if we take a certain kind of transistor and put it into an extremely large electrical field. Laughlin used his Nobel address to argue that the newly discovered phenomenon was strongly emergent. Indeed, he went so far as to say:

> The world is full of things for which one's understanding, i.e. one's ability to predict what will happen in an experience, is degraded by taking the system apart, including most delightfully the standard model of elementary particles itself. I myself have come to suspect that all the important outstanding problems in physics are emergent in nature, including quantum gravity.[29]

Another Nobel Laureate, Philip Anderson, has also influentially argued for further cases of strong emergence in condensed matter physics.[30]

Martin Picard, of the psycho-biology lab at Columbia University, defends the hypothesis that mitochondria in the brain should be understood as social networks rather than reduced to underlying chemistry and physics.[31] On the pan-agentialist view I have outlined, it could turn out that mitochondria have in some way developed motivations beyond the immediate impulses imparted by the wave function, resulting in small deviations from the standard predictions of quantum mechanics. There are also prominent scientists arguing for strong emergence in chemistry and neuroscience.[32]

Much more theorizing would need to be done to flesh out pan-agentialism. How exactly do states of experiential understanding emerge? How does the brain determine which things in the environment are consciously desired? And how do the complex desires of organisms relate to the basic inclinations imparted to particles by the wave function? Answering these questions will involve detailed, philosophically informed, empirical investigation.

Even in the absence of answers to these questions, however, I think we have strong *philosophical* grounds for thinking that any solution to the meaning zombie problem must involve something in the ballpark of pan-agentialism. To see why, we need to go deeper, to uncover the problem within the problem.

The Mystery of Psycho-Physical Harmony

The mystery of psycho-physical harmony is something only a handful of philosophers are talking about at the present moment, but I believe it's going to change the world.[33] It's not an easy problem to get a grip on. That's not because it's especially technical or complicated. It's rather because it demands an explanation of something that's such a mundane, obvious fact of our existence that it feels like it doesn't need explaining. In the same way, it probably wouldn't occur to intelligent fish that the existence of water needs explaining, given that they're swimming in it all the time.

The challenge is to explain why consciousness and behaviour tend to align in a rationally appropriate way. The most obvious examples involve pleasure and pain. When something hurts, that gives you a reason to avoid it. And, in general, humans and other animals respond to that reason by avoiding things that hurt them. Likewise, when something feels good, that gives you a reason to pursue it; in general, humans and other animals respond to that reason by pursuing things that make them feel good. Obviously these reasons can be balanced out by other considerations: I persist through the pain of writing this book because I judge the outcome to be worth the suffering. But the conjoining of pain to avoidance behaviour and pleasure to attraction behaviour is too frequent to be coincidental.

More generally, we tend to respond in a rationally appropriate way to our experiential understanding of the world around us, or at least in a way that makes some kind of rational sense. If my experiential understanding suggests to me that there is a chair in front of me, I tend to behave as though there's a chair in front of me, for example, by sitting on it if I feel like a rest. Of course, humans make all sorts of mistakes and are subject to many silly confusions, but, on the whole, we respond to the reasons our desires give rise to in the light of our understanding of the world. In other words, we respond rationally to the character of our experience.

This all seems so totally obvious as to not need explaining. Of course if something hurts you're going to avoid it! However, when one very carefully reflects on the matter, in the strange, abstract way that philosophers are trained to do, it becomes apparent that the rationally harmonious alignment between consciousness and behaviour *does* need explaining. At least it needs explaining if, as a majority of philosophers believe, *there's no logical connection between consciousness and behaviour.*[34] To explain what I mean by this, let me introduce you to another of my imaginary friends: Inverted Ian.

Inverted Ian feels a strong conscious aversion to burgers and a strong conscious desire to be hacked into little pieces. However, Inverted Ian lives in a parallel universe, and the laws of nature in Ian's universe are different to the laws of nature in our world. In Ian's world, a conscious aversion to something causes you to seek it out, whilst a conscious attraction to something causes you to try to avoid it. As a result, Inverted Ian ends up behaving just like a person does in our world if they have an immense appetite for burgers and a normal fear of being chopped up.[35]

There is obviously something absurd about Inverted Ian. But what is it? There doesn't seem to be anything incoherent about someone's conscious attraction to X causing them to avoid X, or someone's conscious aversion to Y causing them to seek out Y. Conscious experiences are defined in terms of how they feel not what they do, and so there is no contradiction in supposing that conscious experiences could do very different things in different possible universes. Indeed, a significant minority of philosophers have defended *epiphenomenalism*: the view that conscious experience exists but has no impact on the physical body or brain, in something like the way the steam produced by the Hogwarts Express has no impact on the train.[36] If there were some logical connections between conscious experiences and their behavioural effects, then epiphenomenalism would be an incoherent position. I have no doubt that epiphenomenalism is wrong, but it's not contradictory in the way that a square circle is contradictory.

What's absurd with Inverted Ian is that he behaves in a way that is *rationally inappropriate*, given his experience. If you strongly dislike something, it is rational (all things being equal) to avoid it; if you really like something, it is rational (all things being equal) to pursue it. It is such an obvious mundane fact of life that humans and non-human animals respond rationally in at least this basic sense, that we take it entirely for granted; we take it as not needing explanation. But it does need explanation. If there is no logical connection between our experience and the behaviour that results from it, why would it be that conscious experience and behaviour line up in a rationally appropriate way? If we just live in a fundamentally meaningless universe where what stuff does is determined by mathematical laws of nature, why should the behaviour that conscious states produce respect norms of rationality?

Inverted Ian is an especially obtuse example. It's not just that his experience and his behaviour are *not* paired up in a rationally appropriate way; the way they *are* paired up is the very opposite of rationally appropriate. If we did not live in a pan-agentialist world, or more generally in a world where things are somehow set up to ensure that the behaviour of a physical system is rationally appropriate (relative to its conscious experience), then physical systems would likely respond to their experience in a way that had nothing to do with norms of rationality. And if it is unlikely that a physical system would respond rationally to its experiential understanding of the world, thereby surviving well, then evolution has no motivation to endow physical systems with experiential understanding in the first place. Without some inherent push for matter to respond rationally to the character of its experience, our universe would almost certainly be populated by meaning zombies.

We could put the line of argument as follows:

- Experiential understanding is good for survival only if physical systems are likely to respond to it in a rational way, i.e. by pursuing what they desire in the light of what they believe.
- But in the absence of something about our universe that ensures (or makes it likely) that behaviour and experience are paired up in a rationally appropriate way, this is highly unlikely. Given that there's no logical connection between experience and behaviour, it'd be an incredible fluke if the behaviour that resulted from conscious experiences just happened to be rationally appropriate relative to the character of the experience, e.g. conscious aversions causing avoidance behaviour.
- Therefore, an evolutionary explanation of the emergence of experiential understanding depends on there being something about our universe that ensures (or makes it likely) that behaviour and experience are paired up in a rationally appropriate way.

George Orwell once said that 'to see what is in front of one's nose needs a constant struggle.' This is particularly apt when it comes to the puzzle of psycho-physical harmony.[37] It's a really hard problem to get—it just seems so totally obvious that, say, conscious aversions will cause avoidance behaviour. If you feel repulsed by the idea of eating excrement, then clearly you're not going to do it. Of course it's true in the actual world that conscious aversion generally leads to avoidance behaviour, but we need to spend time reflecting very carefully—the skill of the philosopher—on our sense that this *has to be so*. I think we feel this sense of inevitability because it's so obvious that avoidance is the *rational thing to do* when you feel an aversion to something. But if the fundamental causal principles governing our universe have no interest in rationality, then there should be no expectation that this rational connection between aversion and avoidance will be respected. Why shouldn't feeling repulsed by eating excrement lead you to eat it, or to do something totally irrelevant like painting everything blue?

The most common response I've had when explaining the mystery of psycho-physical harmony is: 'natural selection solves it.' But if you think natural selection is the answer to psycho-physical harmony, you haven't understood the problem. This is because any evolutionary explanation of the character of our conscious experience *already assumes* a solution to the problem of psycho-physical harmony, and therefore cannot also solve the problem, on pain of circularity. Natural selection will be motivated to make us feel repulsed by eating excrement only if we are generally going to *respond rationally* to that conscious aversion (by avoiding eating excrement), but the fact that we tend to *respond rationally* to our experiences is the very thing we are trying to explain.

Could experiential understanding be a *spandrel*, that is to say, something thrown up by evolution as a by-product of some other characteristic rather than something which is itself good for survival? I doubt it. It'd be an extraordinary fluke if some characteristic selected for survival just happened to throw up experiential understanding perfectly matched to our functional understanding. We need to give an evolutionary account of the emergence of experiential understanding. But this is possible only if we assume that there is something about the universe that ensures, or makes likely, psycho-physical harmony. Pan-agentialism seems to me the simplest way to do this.[38]

We are happy to believe wacky-sounding theories—that time slows down at high speeds, for example—provided they have hard data to support them. Like special relativity, pan-agentialism *does* have hard data supporting it: the reality of experiential understanding. It's just that the hard data is provided not by experiments but by attending to one's own experience.

The Purpose in Consciousness

Before the scientific revolution, our picture of the universe was dominated by the thought of the ancient Greek thinker Aristotle. Aristotle's universe was filled with purpose. The four elements—earth, air, water, fire—each aimed to get back to their natural place in the order of things: earth fell downwards trying to get to its natural place in the centre for the universe (which Aristotle believed to be in the centre of the planet Earth), fire rose upwards to try to reach the heavens. Organisms, for Aristotle, were not simply mechanisms, but were rather essentially defined in terms of their goal-directed nature. A human being was a rational animal, and as such flourished when exhibiting rationality. Aristotle did not support a subjectivist view according to which what's good for an individual is determined by that individual's personal desires. Rather, the rational nature of all human beings ensured that the life of reason is what is good for us, whether we like it or not. For Aristotle, irrationality is bad for human beings, in the way lack of sunlight is bad for a plant. That's just the kind of creature we are.

The scientific revolution replaced purpose with mechanism. But perhaps purpose was still lurking under the surface, masquerading as mechanism. This is the pan-agentialist view: the stuff of the world is *rational stuff*. Even when behaving in a predictable way, its behaviour is the result of a rational impulse, albeit of a very crude kind. As matter evolves into complex forms, more and more the potential for rational thought and action begins to flower, blossoming in the reflective consciousness of a human being able to discern and respond to practical and theoretical reasons. This is not just matter changing, but matter *maturing*, coming to a greater realization of its inherent rational nature.

The pan-agentialist world is, by definition, a world that embodies purpose. Crucially, however, purpose in this sense does not imply design. Aristotle did believe in some kind of first cause—an 'unmoved mover'— but not a beneficent designer who had crafted the purposes of things. In Aristotle's worldview, things just had goal-directed natures, regardless of their origins. Likewise, if matter, in its fundamental nature, is directed towards reason, then matter has a goal-directed purpose or nature regardless of whether or not it was designed. It is in this sense that consciousness points to purpose, as an essential component of the best explanation of the emergence of experiential understanding.

Furthermore, if we take pan-agentialism as established, and consider it in relation to the fine-tuning of physics, these commitments mutually reinforce the reality of purpose which each individually implies. Rational matter without experiential understanding is only able to realize the most basic rational response: do what you feel like doing. It is only when, through evolution, experiential understanding is selected for, that rational matter achieves a much greater realization of its nature: the capacity to respond to what things are and what things mean. As far as we can tell, in the vast majority of the possible universes generated by varying the values of the fine-tuned constants in physics, the chemical complexity required for evolution would not have been possible. Rational matter would have been stuck in a world in which it was unable to achieve a higher realization of its nature. In other words, the laws of physics are fine-tuned not just for life but for the possibility of rational matter achieving a higher realization of its nature.

Conversely, if the laws of physics had been fine-tuned for life but the universe did not contain rational matter, i.e. matter inherently disposed to respond in a rationally appropriate manner to the character of its conscious experience, it's highly unlikely that experiential understanding would have evolved; the Earth would almost certainly be populated by meaning zombies.[39] As explained above, experiential understanding of reality is helpful for survival only if physical systems are able to respond rationally to the character of their conscious experience.

Fine-tuning and rational matter need each other to produce creatures that can understand and respond to what things are and mean. Without fine-tuning, rational matter would be unable to evolve into complex organisms which are responsive to their environment, a pre-condition for the emergence of experiential understanding. Without rational matter, even if matter evolved into complex survival mechanisms, those mechanisms— if conscious at all—would have meaningless experience: they would be meaning zombies.

In other words, fine-tuning and rational matter fit together like a key fits the lock it was made for. This is unlikely to be a coincidence. More likely, the constants are as they are in order to allow rational matter to blossom as intelligent life (I leave it to the reader to fill in this final Bayesian argument for herself, as a piece of homework). But how could there be such purpose inherent in the universe? The most familiar answer, at least in the West, is *the Omni-God*: an all-knowing, all-powerful, perfectly good creator of the universe. Unfortunately, this answer has serious problems, as we shall discover in Chapter 4.

Choice Point: Readers convinced of the argument of this chapter may proceed directly to Chapter 4, where we begin to theorize about the source of cosmic purpose. For those with objections, stick around to dig deeper.

Digging Deeper

The 'Digging Deeper' section of this chapter mainly consists of responses to objections. But I'll begin by briefly formulating a Bayesian version of the argument for pan-agentialism.

A Bayesian Argument for Pan-Agentialism

In Chapter 2 we introduced the following principle, which can be derived from Bayes' theorem:

> *The Comparative Likelihood Principle*: If the evidence is more likely assuming theory A is true than it is assuming theory B is true, then the evidence supports theory A over theory B.

I used this form of the Likelihood Principle in Chapter 2 to argue that the fine-tuning of physics for life supports the Value-Selection Hypothesis over the Cosmic Fluke Hypothesis, as the evidence of fine-tuning is far more likely on the former hypothesis than on the latter. By exactly the same principle, the existence of experiential understanding supports pan-agentialism over our standard scientific worldview, as it is much more likely that experiential understanding would have evolved on the former hypothesis than on the latter.

On the pan-agentialist view, physical systems respond rationally to the character of their experience; assuming this view, systems that happen to develop a little experiential understanding are likely to survive better, as they will respond rationally to their experiential understanding and thus negotiate the world more successfully. According to pan-agentialism, therefore, it is not unlikely that experiential understanding would be selected for. On the more standard *arational* view of nature, by which I mean the view that the fundamental causal principles governing our universe have no interest in rationality, it is vanishingly unlikely that any creature that happened to develop some experiential understanding would respond rationally to that experience. Thus, if very basic forms of experiential understanding happened to emerge by chance, there is no reason to think they would be helpful for survival, and hence no reason to expect complex forms of experiential understanding to evolve.

In other words, on our standard way of thinking about evidential support, the existence of experiential understanding strongly confirms pan-agentialism over an arational view of nature. Many of the objections to the Bayesian argument of Chapter 7 would also be potential objections to the Bayesian argument of this chapter, so I would invite the reader to look at the 'Digging Deeper' section of Chapter 7 to check out the objections and responses there, in addition to those considered below.

Objection 1: The Idea of 'Free Will' Makes No Sense

Pan-agentialism is not necessarily committed to strong free will—what philosophers call 'libertarian free will.' But I have expressed a preference for forms of it that do involve libertarian free will.

In a recent study of the philosophical views of Anglophone philosophers, only 18.83 per cent professed to accept or lean towards libertarian free will. Having said that, libertarian free will was more popular than *no* free will, which only 11.21 per cent supported (this is interesting, as one can get the impression in the world of online philosophy that the non-existence of free will is the only intellectually credible position). By far the most popular option among Anglophone philosophers, supported by 59.16 per cent (quite rare to get majority support for something in philosophy!) is *compatibilism*.[40]

Compatibilists want to have their cake and eat it. They believe *both* that our choices are determined by prior causes and *also* that we are free.[41] At first it sounds like a contradiction. If there was a chain of causation stretching back to the Big Bang which rendered it inevitable that I would act as I did, and so I could not have done otherwise, in what sense was my choice free? For compatibilists, what's important for freedom is not whether your choices are determined, but whether or not your actions flow from your desires. People locked up in prison are not doing what they want. They would like to see their family, take vacations, and go for long walks by the sea, but they don't do these things because they are confined behind prison walls. Similarly, people with very little money don't do what they want; they are severely confined by their lack of financial options. In contrast, well-financed, non-imprisoned people tend to do what they want to do. For the compatibilist, this is just what freedom is: doing what you want to do. Of course, if everything about us is determined by prior causes, then what we want to do is also determined by prior causes. My former PhD supervisor Galen Strawson famously argued that we are not free because we are not free to shape who we are.[42] That sounds wrong at first. Surely I can freely decide, say, to get fit. But your decision to get fit presumably resulted from a feeling of motivation. For Strawson, the question is what caused that feeling of motivation. Maybe you caused it, by an earlier decision to push yourself. It is familiar to all of us how regularly pushing ourselves makes it easier to do so in future. But what caused that earlier decision to push yourself? Presumably some motivation you had then, which was caused by some earlier motivation, which was caused by some earlier motivation... and so on, until eventually, we get back to something you did not cause, such as the influences of those around you or your actions as a baby. Ultimately, thinks Strawson, we do not decide how we will be, and thus ultimately, we are neither free nor responsible for our actions.

I don't agree with my erstwhile teacher on this. In fact, I think this argument assumes what it's trying to prove. At each stage of the imagined regress stretching back in time, Strawson assumes that choices are determined either by the feelings or motivations of the person, or by things outside of the person. But this already assumes that libertarian freedom doesn't exist. For the believer in libertarian free will, our free choices are not determined by anything, not even our own feelings or motivations. Or rather, the only thing that determines the choice is the chooser herself.

If the chooser's choice is not determined by any of her feelings or motivations, doesn't that mean it's totally random? I don't think so, because in choosing the person will be voluntarily responding to her conscious inclinations or to rational considerations regarding how best to act. Let's take a classic case involving wrestling with oneself about what to do. Susan is happily married. Away at a work conference, she is engaged in what seems at first to be harmless flirting at the hotel bar, but, after a few drinks, starts to turn more intimate. Upon returning to her hotel room, Susan finds the number of the woman she was flirting with in her pocket with a message, 'Call me. I want you.' Susan feels a rush of desire. She wants this woman so badly, and knows that if she doesn't give in to this temptation tonight, she'll never see her again. But Susan also knows that this is something she shouldn't do. She has promised fidelity to her wife and knows how much that matters to her wife. Susan is also not a very good liar and knows there's a good chance her wife would somehow find out about the affair, and that such a revelation would likely destroy their very happy marriage. For both moral and prudential reasons, there is no doubt about it: this is certainly not a good idea. But, God, she really wants to.

Susan is torn between two very different kinds of consideration:

A Value Consideration: In terms of her value judgements, Susan knows that having an affair would be a very bad idea.

A Conscious Yearning: Susan feels a very strong desire to have an affair.

On a libertarian conception of free will, no prior cause will determine which way Susan will go. It is up to her to decide. But when she does decide, her choice won't just be a random, spontaneous happening, like the random decay of a radioactive isotope.[43] It will be a conscious decision to respond to one of the above two rational considerations, and thus there is a logic to it which distinguishes it from a random, senseless occurrence.

But what determined that Susan ended up going one way or the other? Nothing, according to the libertarian. Susan decided, say, not to have the affair (let's give our story a happy ending). But there is no deeper explanation of the fact that she decided to go one way or the other. She just did. As Ludwig Wittgenstein said, explanations have to end somewhere.[44]

Arguments against free will tend to blur together two very different arguments:

Argument 1:   If Susan's decision lacked a prior cause, then it was random.

Argument 2:   If Susan's decision lacked a prior cause, there was no explanation of why Susan decided one way rather than the other.

I've found in arguing about this (as I've already said, I spend too long on Twitter...), when you respond to Argument 1—by explaining that Susan's choice is not random because it essentially involves responding to rational considerations—people turn to Argument 2. And then when you respond to Argument 2—by pointing out that explanations have to end somewhere—people jump back to Argument 1. But once these two arguments are clearly distinguished, it can be seen that there is no good argument against the coherence of libertarian free will.

Of course, just because free will is coherent, it doesn't mean it exists. Unicorns are coherent, and my 6-year-old daughter is convinced they exist deep in forests, hidden from humans due to their shyness. But unicorns don't exist, and it could be that libertarian free will doesn't either. This is ultimately an empirical question, but, as I argued on pp. 69–70, we are a long way from knowing enough about the brain to know if everything that goes on in it is reducible to underlying chemistry and physics. In the near term, whether or not we have reason to believe in free will depend on philosophical considerations of the kind discussed in this chapter.

Objection 2: The Idea of a 'Meaning Zombie' Doesn't Make Sense

The 'meaning zombie problem' is the challenge of explaining why we didn't evolve as meaning zombies. The most straightforward way of avoiding this problem is to deny that the concept of a meaning zombie is coherent—that it even makes sense. For the proponent of this objection, meaning zombies seem to make sense when you first think about it, but on reflection they turn out to be as incoherent as square circles.

If the concept of a meaning zombie is incoherent, this is presumably because there is a logical connection between *experiential understanding* (the conscious grasp of what things are and mean) and *functional under-standing* (the way in which understanding is exhibited in behaviour or in the information processing of the brain). And if there is a logical connec-tion between experiential understanding and functional understanding, this is presumably because all we *mean* when we say that someone 'experi-entially understands' such and such is that they 'functionally understand' such and such. This seems to me a wildly implausible claim about what our words mean. My grip on the meaning of 'experiential understanding' comes from attending to the character of my own experience, whereas the concept of 'functional understanding' is a more theoretical notion rooted in cogni-tive science. It would be odd if these terms that originate in very different contexts turn out to be synonymous.

Maybe the thought is that the general idea of a 'philosophical zombie' doesn't make sense. Whether that's right hangs and falls on whether talk of 'feelings and experiences' is just a way of talking about behaviour (either external behaviour of the organism or the internal behaviour of bits of the brain). All I mean when I say 'there's a party at Susan's place' is that there are people partying at Susan's, and hence the idea of people partying in all of the normal ways at Susan's in the absence of a party makes no sense. But when I say, 'Susan is feeling anxious,' I'm not meaning to make a claim about how she's disposed to behave, or how her internal parts are disposed to behave. And because I'm not making a claim about Susan's behaviour (or that of her parts), there's no contradiction in claiming that everything about Susan's behaviour points to anxiety even though she doesn't actually feel anxious. And the same goes for all of Susan's feelings and experiences. In other words, there's no contradiction in supposing that Susan is a zombie.

Remember we're not discussing whether it's *reasonable* to think that Susan is a zombie or a meaning zombie, simply whether it's *coherent*, in the sense of not involving a contradiction. The contents of the Harry Potter books are—on the whole—coherent even though they depict things that could never happen in our universe. Nobody thinks zombies or meaning zombies are real but the *idea* of zombies and meaning zombies is at least coherent.[45]

Objection 3: Meaning Zombies Are Impossible

A more nuanced response to the meaning zombie problem is to argue that meaning zombies are *coherent but impossible*. The words 'water' and 'H_2O' don't mean the same thing, but science has shown that water just is H_2O. Similarly, although 'experiential understanding' and 'functional understanding' don't mean the same thing, perhaps it could be argued that they are nonetheless identical.

I spend the first half of my first book *Consciousness and Fundamental Reality* arguing against these kinds of identities between experiential and physical or functional states. But actually, all that's required to get the meaning zombie problem going is the *coherence* of meaning zombies, not their genuine possibility. As we discussed in the 'Digging Deeper' section of Chapter 2, Bayesian possibilities are not objective possibilities but are rather rooted in what it is coherent to believe. Even though water is identical with H_2O, in Bayesian reasoning we could still consider coherent scenarios in which water turns out not to be H_2O (indeed, we would need to do so, for example, when evaluating the evidential support for the hypothesis that water is H_2O). Similarly, even if meaning zombies are impossible—perhaps because experiential understanding is identical with functional understanding—so long as meaning zombies are conceptually coherent, we can wield the likelihood principle, as we did above, to make the case for pan-agentialism over the arational view of nature.

Objection 4: Experiential Understanding Doesn't Exist

It might surprise readers that probably the most controversial bit of the meaning zombie thought experiment is the reality of experiential understanding itself. In the latter half of the 20th century, the consensus position in Anglophone philosophy was that thought and understanding were nothing to do with consciousness. It was taken for granted that things like thought or understanding could be accounted for in terms of behaviour or behavioural functioning, or perhaps in terms of causal relationships with things in the environment. This is still probably the view of the majority, although significant dissent has emerged in the last fifteen years or so.[46]

In my view, the reality of experiential understanding is as evident as the reality of pain. I know that pain hurts because I'm directly aware of its hurty character. Similarly, I know that the character of my visual experience of my crying child represents her sadness, or my auditory experience of someone shouting 'Get out, the building is on fire!' represents the meanings of those words, because I am directly aware of the understanding-involving character of those experiences. One way to bring this out is to go through the process of Cartesian doubt, doubting the reality of your body and your brain and the external world around you, until you end up conceiving of yourself as a pure conscious being: a thing with no properties other than the having of a certain form of conscious experience. When I do this, the thing I end up conceiving of is a thing that has understanding of the world it thinks it's living in. My purely conscious twin thinks it's sitting with a laptop typing words which it experiences as having certain meanings, for example. To put it another way, any possible creature that had the same conscious experience as me right now—i.e. any creature such that what's it's like to be that creature is exactly the same as what it's like to be me right now—would also think it's sitting with a laptop typing words with certain specific meanings. It follows that there is a necessary connection between my conscious experience and my understanding.

It's strange that this debate persists. It's a debate about the character of human experience. Given that we are humans, this ought to be something we can settle just by attending to the character of our experience. However, when some philosophers attend to their experience, they claim to find the reality of experiential understanding totally evident; when other philosophers attend to their experience, they claim to find just sensory qualities (colours, sounds, smells, tastes) or raw sensations (pain, itchiness).[47] Even for those in the latter camp, so long as the *awfulness* of pain is part of the conscious experience, then that's enough to raise the question of why awful experiences tend to produce avoidance behaviour (see the above section 'The Mystery of Psycho-Physical Harmony'), a question which I believe can only be answered by something like pan-agentialism.

For those readers who don't believe in experiential understanding, and don't think the awfulness of pain is part of the character of the pain experience itself, I concede that you will be entirely unmotivated to accept my argument. I suspect, however, that there won't be many.

4

Why the Omni-God Probably
Doesn't Exist

In 1986, a piece in the *Detroit Free Press* reported how a 5-year-old girl had been found raped, severely beaten, and strangled to death.[1] The perpetrator was the drunk and jealous boyfriend of the mother. If the traditional beliefs of any one of the Abrahamic faiths—Judaism, Christianity, and Islam—are correct, there was someone who witnessed this horrific event as it was happening, was able to stop it, and chose not to. If there is an all-knowing, all-powerful, and perfectly loving God—an 'Omni-God'—this Supreme Being looks on and chooses not to intervene to stop the terrible suffering and oppression we find in the world.

It can seem flippant to raise such a horrific event in the context of an abstract intellectual discussion. Whenever I'm teaching the 'problem of evil'—the challenge of reconciling the existence of the Omni-God with the evil and suffering we find in the world—I emphasize to my students that I'm not treating these events lightly, and appreciate that many may find hearing them and thinking about them difficult. Unfortunately, this is what we need to do to seriously engage with one of the biggest questions of philosophy: Does God exist?

Why would a loving God allow terrible pain and suffering to occur? Attempts by philosophers and theologians to explain what God's reasons might be are known as 'theodicies.' The most well-known is the *free will theodicy*, which traces back to Saint Augustine in the 4th century. Human freedom, it is claimed, is a great good, which is why the Omni-God has created us with free will. But if you're going to give people free will, then you open up the possibility that certain people will use their free will to commit terrible atrocities, such as that with which we began this chapter. The only alternative is to make robots programmed to always do good. According to the free will theodicy, the Omni-God decided that the good of having free creatures outweighed the bad that would result from the misuse of their freedom.

One concern with the free will theodicy is that it assumes the very strong form of free will discussed in Chapter 3: libertarian free will. For those who think libertarian free will is incoherent or at least non-existent, this theodicy is not going to be an option.[2] But given that I am myself sympathetic to the existence of libertarian free will, and defended the coherence of the concept of libertarian free will in the 'Digging Deeper' section of Chapter 3, I shall set this objection aside.

Even assuming the existence of libertarian free will, at best the free will theodicy can account for the terrible things human beings do to each other. What is left unexplained is the terrible suffering caused by the natural world. There are over ten million views on YouTube of the atheist actor and comedian Stephen Fry answering the question of what he'd say to the Omni-God if he ended up meeting Her on the other side of the grave:

> I think I'd say: Bone cancer in children...what's that about? How dare you? How dare you create a world in which there is such misery that is not our fault? It's not right, it's utterly, utterly, evil. Why should I respect a capricious, mean-minded, stupid God, who creates a world which is so full of injustice and pain?...Yes the world is very splendid, but it also has in it insects whose whole life cycle is to burrow into the eyes of children and make them blind—they eat outwards from the eyes! Why?! Why did you do that to us?? You could easily have made a creation in which that didn't exist. It is simply not acceptable!

There are many, many other examples of horrific suffering in nature. The philosopher Felipe Leon compiled a long list in making the case against the Omni-God, including the agony caused by the North American short-tailed shrew, that paralyses its prey then eats them alive over several days, the female mantis that eats its fully conscious lover during intercourse, and parasitic wasps that lay eggs inside their victims, which subsequently hatch and consume their host from the inside out.[3]

Setting aside these especially grim examples, horrific suffering is built into the very process that created us: natural selection. Why would an all-powerful being choose to bring us into existence through such a gruesome, long-winded, tortuous process as a game of 'survival of the fittest'? Why not just create a person by breathing spirit into dust, as we find depicted in Genesis? And however we're created, why give us bodies that age, get sick, and fall apart so easily? Why not instead create immortal, spiritual beings to spend eternity in loving union with God and with each other?

One of the most celebrated defenders of the Omni-God in contemporary philosophy is Richard Swinburne. Perhaps Swinburne's most important innovation is bringing the Bayesian reasoning we discussed in the 'Digger Deeper' section of Chapter 2 into the philosophy of religion, introducing a rigorous framework to previously undisciplined discussions. He has also tried to extend the free will theodicy to cover suffering that is not of human creation.[4]

Swinburne does this in two ways. First, he argues that nature's cruelty is required in order to *expand the range of choices available to us*. Without the tsunami in the Indian Ocean in 2004, which killed an estimated 227,898 people in fourteen countries, there would be no opportunity for people unaffected by the disaster to choose whether to show compassion and assist the victims—whether physically or financially—or to dismiss the tragedy as 'not my problem.' Without the dangers that fill the natural world, there would be little opportunity for people to choose whether or not to be courageous in facing great danger. Whilst Swinburne believes that the Omni-God could have made a world of immortal spirits with all their needs satisfied, such a world would be a kind of Disneyland without the moral seriousness we find in the real world. Strictly speaking we would have free will, and strictly speaking perhaps people could choose to harm each other. But there would be little cause to do so, and the range of choices available to us to 'freely' choose from would be paltry. All created beings would be like the idle rich, their most weighty choices being which form of ecstasy to enjoy today.

Second, Swinburne argues that nature's cruelty is required for us to have knowledge of how to cause good and evil, a prerequisite of moral choice. If nobody had ever been observed to die from consuming a poisonous plant, then nobody would know about the lethal properties of that plant, and hence nobody would have the moral choice of whether to wilfully poison others using it. If nobody had ever tripped and fallen to their death from a great height, then we wouldn't know that giving people a nudge off a cliff will send them to their doom. There needs to be a natural world that works in a regular, mechanical way—sometimes harming and sometimes hindering—in order for humans, through their observations of that world, to learn how they themselves can cause good or ill.

Surely the Omni-God could have created people who are born with intuitive knowledge of how to help and harm each other? After all, the Omni-God can do anything She likes. Swinburne acknowledges this possibility, but argues that such a world would lack the great good of being able to find out about reality through scientific enquiry. We would also not face the serious choice of deciding whether to pursue scientific enquiry—whether, for example, to spend public money investing in medical science or instead to prioritize something else, such as cutting taxes. For Swinburne, a world where we were all born with complete knowledge of how the world works would be too easy, and thus would lack the gravitas of the real world. Without the possibility of scientific enquiry, there would be no Newton or Einstein, just as without the dangers of nature, there would be no Amundsen or Scott.

Personally, I am unconvinced by these proposals. I agree that there are certain goods we find in the real world—compassion, courage, adventure, scientific enquiry—that would not exist in a more perfect world. But it seems to me to massively reduce the value of these goods if they were brought about through artificially engineering challenges and difficulties. Consider the following thought experiment:

Great Discovery: After ten years of painstaking research, I finally discovered the double helix structure of DNA. It was the proudest moment of my life. Shortly after the words 'Game Over' flashed in front of me, and I woke up in another world. After a debriefing session, I found out I live in a Utopia where all scientific discoveries have been made and all human needs met. Out of boredom, humans in this society regularly choose to have their memories wiped so that they can be fully immersed into a virtual world in which society is ignorant of certain important scientific truths. In this way, people can then enjoy the challenge of trying to 'discover' these truths for themselves.

If I really did manage to work out the structure of DNA through my own efforts, then there is real value in that achievement. If the future really is that boring, such titillation may be the only alternative to suicide. However, upon leaving the game, I would no doubt be crestfallen to discover that I had not advanced knowledge as I had thought. Great scientific discovery has no more worth than solving a tricky puzzle if the facts are already known but have been deliberately withheld for the sake of a challenge. But on Swinburne's view, this is exactly the reality. In any scientific discovery, the facts are already known—by God—and the only reason we don't also know them is to allow us the fun of discovering them. To my mind, this makes science a trivial game.

In the case of scientific discovery, the value of the good in question is merely lessened. There is a more serious ethical concern when the challenges are artificially created through the deliberate endangerment of life and limb. Here's another thought experiment:

Heroic Rescue: I'm walking home from work one day and happen upon a burning building in which a baby can be heard screaming from a top floor window. Without a second's thought, I leap into action: shimmying up a lamppost, leaping onto the rooftop, and lowering myself into the apartment through a window. I grab the baby and bravely make my way back to the rooftop and then to the ground. Crowds of onlookers break into cheers and applause as I give the baby to her grateful mother. At that moment, the hidden camera crew emerges, and the host comes up to warmly ask me how I feel about winning 100 points on the 'Who Dares?' gameshow. It turns out the whole thing was a set-up; the baby was deliberately put in danger to test whether or not the next passer-by would show the courage needed to rescue it.

No doubt I displayed great skill and courage in the rescue. And if the child's life really was in danger, then I managed to prevent a tragedy. But there is something perverse about a world in which lives are put in danger just for the sake of facilitating daring rescues. The whole thing is reminiscent of the brilliant but terrifying Korean drama 'Squid Game,' in which contestants risk their lives in lethal games of chance and skill. Even if courage and compassion are great goods, it is immoral to endanger people merely for the sake of allowing others to show courage and compassion.

To be fair, although the above thought experiments are vivid, there are some disanalogies to the case of Swinburne's Omni-God. Unlike in my 'Heroic Rescue' thought experiment, the Omni-God doesn't put babies in danger to titillate a TV audience (although you never know...maybe it's primetime entertainment for the angels...). And whilst in the 'Great Discovery' thought experiment, somebody has already discovered the structure of DNA, in Swinburne's world, although the Omni-God already knew the structure of DNA, nobody *discovered* this before Crick and Watson (the Omni-God knew because She created DNA!). Still, what seems to me most problematic in these examples is how artificial and unnecessary the situation is. In 'Great Discovery,' there is no real need for ignorance about the structure of DNA other than to create the challenge of discovering it. In 'Heroic Rescue,' there is no real need for the baby to be in danger other than to create a situation where someone can choose whether to show courage and compassion. These elements are present in Swinburne's account of the Omni-God's reasons for creating suffering. If no one was seriously harmed by the challenges of life, then we could perhaps understand God's decision to give us lives of adventure as opposed to dull opulence. But the challenges God has given us cause horrific pain and suffering. If Swinburne's explanation of why we suffer is correct, then this divinely dictated drama of human existence is a cruel farce.

To his credit, Swinburne does consider whether God has the right to harm people in this way. He is concerned to show not only that a world with natural evil is better than a world without (because the former contains courage, compassion, scientific discovery, etc.) but also that our creator has the right to use people to make the world better. Suppose a doctor could save the lives of five patients by killing one healthy person and harvesting their organs: giving the heart to one patient, the kidneys to another, etc. Most people think it would be wrong for the doctor to do it, as it would be violating the right to life of the healthy person. Similarly, God surely knew that Her decision to create natural evil would result in terrible suffering to countless numbers. Even if we concede to Swinburne that this decision made the world better, all things considered, did God not violate the rights to life and security of the people harmed by natural evil?

In response, Swinburne argues that 'as author of our being [God] would have rights over us that we do not have over our fellow humans.' To make a case for this, Swinburne argues that a parent has the right to force one of her or his children to suffer 'for the good of his or his brother's soul'; given that the parent is responsible for their child's existence and continued life, they have 'the right to demand something in return.'[5] If a mere human parent has this right to some extent, argues Swinburne, so much greater must God who is the source of all being. On this basis, Swinburne thinks that God does have the right to allow people to suffer in natural disasters, so long as She compensates the victims in the life to come.

It's unfortunate that Swinburne doesn't give an example of what he means when he says that a parent has the right to force their child to suffer 'for the good of his or his brother's soul.' I guess I have the right to force my older daughter to let my younger daughter play with her toys, for example, if the latter is a bit unwell and needs cheering up. But I think this 'right' I have over my daughter is due to the fact that my daughter is only 6 and so is not fully competent to make all of her own decisions, rather than to the fact that I created my daughter. Biological parents do not have more rights over their children than adoptive parents. Perhaps in the future it will be possible to artificially create conscious and autonomous systems. If those systems really are free and self-aware, then they should be afforded the same rights as human beings, and those rights should not be constrained by the desires of their human creators. Likewise, I don't want to sound ungrateful, but I don't accept that my creator would have special rights over me in virtue of having brought me into existence, at least once I'm an autonomous adult capable of making my own decisions.

It therefore seems to me not only that the Omni-God would have no good grounds for allowing natural evil, it would actually be immoral for Her to do so, given the widespread violations of the rights to health and happiness that would result. I call this the 'Cosmic Sin Intuition', which we can define as follows:

The Cosmic Sin Intuition: It would be immoral for an all-powerful being to deliberately create a universe like ours.

I'm a big fan of the dystopian Sci Fi show *Black Mirror*. One of its episodes focuses on the touching love story of Frank and Amy, who meet through a dating app that matches couples for a set period of time. When their time is up and Frank and Amy are then matched with other partners, they yearn to be together again. It slowly dawns on them that neither can remember life before starting to use the app, and that things seem to be conspiring to ensure they keep using it. Eventually, they rebel and escape together, only to find themselves escaping from a simulation, one of 1,000 Frank and Amy simulations of which 998 rebelled. As they emerge, Frank and Amy evaporate as they too are part of the simulation, part of an app in the real world designed to test the dating compatibility of real-world Frank and Amy. At the end of the episode, real-world Frank and Amy check their phones and are told they are a 99.8 per cent match.

Suppose in the future it will be possible to create programs that simulate all the physical complexity of an actual universe, right down to the electrons and quarks. Let us further suppose, as is represented in the *Black Mirror* episode, simulated creatures that emerge in such simulated universes would have all the same feelings and experiences as flesh and blood organisms have, given that the simulations of their brains have all the same detailed structure we find in our own physical brains (we will revisit this assumption in Chapter 5). We can hope that technologically advanced societies will have minimal moral standards, at least having the pretence of being opposed to senseless suffering even if they don't live up to that standard in practice. And, hence, if creatures in simulated worlds really are conscious, then presumably plans to simulate a world containing sentient creatures would have to be approved by some kind of a 'Committee for the Assessment of the Moral Permissibility of Simulated Worlds.'

Imagine a proposal to simulate a universe just like ours, complete with natural disasters and all the pain and suffering involved in the process of natural selection. I don't think there's much chance the committee would approve these plans. The would-be simulator could present to the committee the kinds of arguments Swinburne puts forth:

> But the great suffering and danger present in my simulation ensures that the people living in my simulation will get the opportunity to show great courage and compassion. Furthermore, their ignorance gives them the opportunity to conduct scientific enquiry and uncover the laws governing my simulation!

My hunch is that this would-be simulator would be laughed out of town. A simulator would essentially be an all-powerful creator with respect to a universe they simulated. Hence, in rejecting such plans the committee would for all intents and purposes be signing up to the Cosmic Sin Intuition.

Given the uncertainty regarding the moral character of future technologically advanced societies, it's not clear we can rule out the possibility that we might be living in a computer simulation created by such a society (more on this in Chapter 5...). But when it comes to a God defined as morally perfect, the Cosmic Sin Intuition does cast considerable doubt on Her existence. If it is immoral for an all-powerful being to create a universe like our own, then either our universe lacks a creator, or that creator is not all-powerful and perfectly good. Either way, the Omni-God does not exist.

I want to finish my discussion of Swinburne's argument with a quotation in which he sums up the values that he believes are facilitated by the Omni-God allowing suffering, but which, to my mind, wonderfully captures the perversity of Swinburne's deity:

> Suppose that one less person had been burnt by the Hiroshima atomic bomb. Then there would have been less opportunity for courage and sympathy; one less piece of information about the effects of atomic radiation, less people (relatives of the person burnt) who would have had a strong desire to campaign for nuclear disarmament and against imperialist expansion.[6]

I rest my case, your honour.[7]

Why Does the Universe Exist?

How come I exist? Because my parents decided to have a fourth child and the result was me. How come my parents exist? Because my grandparents decided to have children and took the necessary steps to ensure they did, as did my great-grandparents, and so on back through the generations. How come the human race exists at all? Because we evolved from earlier forms of life, which evolved from earlier forms of life, and so on right back to the common ancestor of all life on Earth. How come life on Earth exists? Answering that question is still a work in progress, but let us suppose for the sake of discussion that scientists one day nail a chemical explanation of how living organisms emerged from inanimate matter. How come the planet Earth exists to produce life? Because gravity clumped gas clouds together into stars and planets. Astonishingly, we can trace this Great Chain of Explanation right back nearly fourteen billion years to the first split second of the universe, when all of the matter and energy of the universe was concentrated in a tiny region.

So how come the universe came to be in the first place? The idea that we can end this grand story of explanation by saying, 'No reason, it just did,' feels wrong. If there is no explanation of why the universe began to exist in the first place, then there is ultimately no explanation of why any of the incredible stuff we see around us—stars, planets, turtles, trees, and orange juice—exists. Any philosophically minded person is likely to feel the pull of the intuition that this just *cannot* be right, whether or not they end up trusting that intuition.

This demand for an ultimate explanation undergirds one of the most influential families of arguments for the existence of God; the great 17th-century philosopher Immanuel Kant dubbed them 'cosmological' arguments for God. Without a creator God, so the argument goes, we have no explanation for why the universe came to be; indeed, no ultimate explanation of anything at all. A common, and understandable, atheist retort is to query why the universe requires an explanation but somehow God gets a free pass. Any halfway decent cosmological argument for God's existence must respond to this point by explaining the crucial difference between God and the universe that means that the former but not the latter cries out for explanation.

There are broadly speaking two ways of trying to explain this difference:

- *The Kalām Cosmological Argument*: For proponents of the Kalām Cosmological Argument, the crucial difference between God and the universe is that the universe *began to exist* whereas God did not.

Things that begin to exist, so the argument goes, require causal explanation, whereas things that are timeless, or have always existed, do not.[8]

- *The Contingency Argument*: For proponents of the contingency argument, the reason the universe cries out for explanation is that it is *contingent*, which means that *it might not have existed*. If something exists but might not have existed, so the argument goes, then we need to give some kind of explanation as to why it does exist. God, in contrast, is not contingent; if God exists at all, then She is a *necessary being*, i.e. a being that *has to exist*.[9]

Cosmological arguments, of either flavour, proceed in two stages: The first stage of the argument is to try to establish that a timeless and/or necessary being has to exist to complete the Great Chain of Explanation. The second stage is to try to show that this timeless and/or necessary being must have the characteristics of the Omni-God: all-knowing, all-powerful, and perfectly good. For what it's worth, I'm pretty sympathetic to the first stage. There does seem to be something rationally intolerable in the Great Chain of Explanation bottoming out in something just popping into existence for no reason, or with something that might not have existed. Perhaps the Great Chain of Explanation consists of an infinite regress of universes, with the end point of each causing the Big Bang of the next. But that would leave us wanting to know why that regress of physical universes has always existed, as opposed to, say, an infinite regress of ghosts—earlier ghosts giving rise to later ghosts—or nothing at all.

To my mind, the point at which these arguments fall apart is when they attempt to demonstrate that the timeless/necessary foundation of being must have the properties of the Omni-God: all-knowing, all-powerful, and perfectly good. If we are rationally compelled to accept the existence of a timeless, necessarily existent entity, why not suppose that that entity is the universe itself, rather than postulating something supernatural outside of the universe?

Proponents of the Kalām Cosmological Argument respond by pressing scientific and philosophical arguments which purport to demonstrate that the universe began to exist. And if the universe has not always existed, then it cannot exist of necessity; necessary things, by definition, cannot fail to exist and so cannot have begun to exist. But, at best, these arguments prove that the universe's existence *as a spatiotemporal entity* is contingent and began at a finite point in the past. A possibility rarely considered in these discussions is that the universe, prior to the Big Bang, existed *in a non-spatiotemporal form*.[10] It's certainly odd to think of our physical universe existing in such a radically different form. However, once we have committed to the existence of a necessary and timeless foundation of existence, we face a choice between two theoretical possibilities:

Option 1: The necessary and timeless foundation of existence is distinct from the universe and brought the universe into being.

Option 2: The necessary and timeless foundation of existence *became* the universe a finite period of time ago, undergoing a radical change analogous—although obviously more extreme—to a caterpillar becoming a butterfly.

Scientists and philosophers try to respect Ockham's razor, the theoretical imperative to postulate as few entities as possible. Option 2 postulates fewer entities than Option 1, and hence, in the absence of further argument, looks to be the better hypothesis.

Of course, this is not the end of the debate (it never is with philosophy!). Believers in the Omni-God have offered various arguments to try to show that the Great Chain of Explanation must bottom out in the Omni-God. We cannot here provide an exhaustive study of all such arguments, but I want to spend a little time exploring one of the more intriguing of recent attempts. Josh Rasmussen argues that the Ultimate Foundation of all of reality cannot involve *arbitrary limits*.[11] To use his colourful example, it is absurd to suppose that the ultimate source of all being and truth is (was?) shaped like a Disney Princess. The 'Disney Princess' hypothesis is absurd because it would cry out for explanation why the Ultimate Foundation has precisely that shape rather than some other, whereas the Ultimate Foundation, by definition, must leave nothing to be explained.

On this basis, Rasmussen argues that the Ultimate Foundation cannot have *a certain amount* of power, knowledge, and goodness, whilst not being *all*-powerful, *all*-knowing, and *perfectly* good. For it would then cry out for explanation why the Ultimate Foundation has precisely that level of power (and not more or less), that level of knowledge, and that level of goodness—and these explanatory demands would undermine its putative status to be the point where explanations come to an end.

Rasmussen's argument is fascinating, and I think it gets something right. Arbitrary limits do seem to cry out for explanation more than maximal properties. Before we knew that light travelled at some specific, finite speed, scientists assumed it travelled infinitely fast, as the latter is a simpler hypothesis, and thus a more natural resting place for explanation.[12] However, I would also want to emphasize that all of these theoretical options are on a continuum: Simpler hypotheses cry out for explanation *less* than more complex hypotheses, but the need to explain never entirely goes away. So long as *something* exists, human curiosity will demand to know *why* that thing exists. Even if a being with maximal perfections is a more natural stopping point for explanation than a Disney Princess, we would still want to know why that maximally perfect exists rather than the even simpler hypothesis of nothing at all existing.[13]

The moral of the story, I think, is that we should try to make our hypotheses as simple as possible. This is already the practice of scientists and philosophers. But we don't go for the *simplest* hypothesis—the hypothesis that nothing exists at all—but rather the simplest hypothesis *consistent with the data*. And the Omni-God hypothesis, even if it is very simple, is not compatible with the data of evil and suffering.

In summary: whilst those who deny cosmic purpose cannot explain cosmological fine-tuning, believers in the Omni-God cannot explain the evil and suffering we find in the world. It's high time we tried to formulate a hypothesis able to account for both. This is the topic of Chapter 5.

Choice Point: If you have heard enough to be motivated to explore Godless ways of making sense of cosmic purpose, you can proceed to Chapter 5. In the slightly more challenging section of this chapter, we will explore the cutting-edge of the academic debate on the problem of evil.

Digging Deeper

The Lord Works in Mysterious Ways

There have been many, many attempts to explain why the Omni-God would allow suffering, but I personally find Swinburne's to be the most thoughtful and plausible. I also find it totally inadequate. However, it's not yet game over for the Omni-God. The cutting-edge discussion of the problem of evil in philosophy departments these days focuses not so much on theodicies but on the view that has become known as 'skeptical theism.'

A 'theist' is someone who believes in the Omni-God. *Skeptical* theists are distinguished by how they respond to the problem of evil. The first thing to note is that skeptical theists tend to agree with atheists that all of the theodicies that philosophers and theologians have come up with down the ages are inadequate. Here's a nice quote from skeptical theist Alvin Plantinga:

we cannot see *why* our world, with all its ills, would be better than others we think we can imagine, or *what*, in any detail, is God's reason for permitting a given specific and appalling evil. Not only can we not see this, we can't think of any very good possibilities. And here I must say that most attempts to explain *why* God permits evil—*theodicies*, as we may call them—strike me as tepid, shallow and ultimately frivolous.[14]

You might be forgiven for thinking this was penned by Richard Dawkins, when in fact Plantinga is a conservative Christian. Plantinga is convinced that the Omni-God exists, and has perfectly good reasons for allowing pain and suffering. We just have no idea what these reasons are.

On first hearing, skeptical theism can sound desperate. If our best minds have not come up with any plausible reason why an all-powerful, loving God might have created a world like this, doesn't that give us at least a little reason to think that probably there is no reason, and hence that probably the Omni-God doesn't exist? However, skeptical theists respond that, if the Omni-God does exist, then She is vastly superior to us in knowledge and understanding, and hence we should not *expect* to understand the Omni-God's reasons for creating the world. As Saint Paul puts it in the bible:

Oh, the depth of the riches of the wisdom and knowledge of God! How unsearchable his judgments, and his paths beyond tracing out! Who has known the mind of the Lord?[15]

Plantinga has a pithier line:

Why suppose that if God does have a reason for permitting evil, the theist would be the first to know?[16]

Skeptical theists have come up with various analogies aimed at showing that any inference from 'we can't think of a reason why the Omni-God would allow suffering' to 'there probably is no reason that could justify the Omni-God allowing suffering' is hasty and ill-advised:

The Chess Analogy: You're a chess beginner playing a grandmaster. You can't understand the reason for their last move. But it would obviously be ill-advised to infer from this that there is no good reason for the move. Given that you're playing chess with someone vastly more advanced at the game than yourself, you shouldn't *expect* to discern the reasons for their moves.

The Physics Analogy: You're a first-year physics undergraduate struggling to understand a book on quantum theory. You can't for the life of you understand the reason the author draws a certain conclusion. Would it be rational to infer that there is no reason? Of course not! Given the vast difference in knowledge of the subject between you and the author, you should not expect to understand their reasoning.[17]

The difference in cognitive capacity between us and God is presumably many, many times greater than between a chess grandmaster and a beginner, or between Einstein and a physics undergraduate. Thus, if it is ill-advised to infer from 'I can't see a reason to do X' to 'there probably is no good reason to do X' in these cases, then it is presumably ill-advised—to an even greater degree—to infer from 'I can't see what reason an Omni-God would have to allow suffering' to 'there probably is no good reason for an Omni-God to allow suffering.'

I'm not sure these analogies work. It's important to emphasize that skeptical theists are trying to show that the evil and suffering we find in the world is *no evidence whatsoever* against the existence of the Omni-God. To what extent do these analogies show that? In the set-up, we *know* we are playing a chess grandmaster or reading a text written by a knowledgeable physicist. Hence, these analogies are appropriate only if we *know* the world was created by a supreme being. But suppose instead you're playing chess online, and you don't know whether you're playing a grandmaster or a bot that is randomly generating moves, and the moves look really bad. In that case, it would be more rational to infer that you're playing the bot. By analogy, when we look at the world and find suffering that seems to be arbitrary and meaningless, the rational inference to draw is that we are seeing the results of arbitrary chance rather than design.

Setting aside the analogies, skeptical theist Stephen Wykstra came up with a principle to adjudicate between when it is and when it isn't okay to infer from 'I can't see an X' to 'there probably isn't an X,' in order to assess whether the move from 'I can't see what reasons could justify the Omni-God allowing suffering' to 'there probably are no reasons that could justify the Omni-God allowing suffering' is a rationally acceptable inference. The principle has been dubbed the 'CORNEA' principle (an acronym of

Condition Of ReasoNable Epistemic Access), and a very rough version of it is as follows:

CORNEA: To judge whether it's rational to move from 'I can't see an X' to 'there probably isn't an X,' ask yourself the following question: 'If there *were* an X, then would I see it?' If the answer is 'yes', then it's rational to make the inference. If the answer is 'no', then it isn't.[18]

To illustrate the principle in action, suppose you're on the top of a skyscraper looking down at the ground hundreds of metres below. You can't see any ants on the pavement. Is it rational for you to infer from this that there are no ants on the pavement beneath the skyscraper? Of course not! CORNEA makes sense of this: if there are ants on the pavement below, you ain't gonna see them, and hence it would *not* be rational to infer that there are no ants from the fact that you can't see any ants. In contrast, suppose you're sitting in a room with no obvious hiding places and you can't see an elephant. In this case, you certainly are justified in inferring that there probably isn't an elephant, and again CORNEA accommodates this. If there were an elephant, you'd most likely spot it, and hence it's rational to infer from the fact that you can't see an elephant, that there probably isn't one there.

Returning to the case of the Omni-God, we want to know whether it's rational to infer from 'I can't see a reason that would justify the Omni-God allowing suffering' to 'there probably isn't a reason that would justify the Omni-God allowing suffering.' In line with CORNEA, we need to ask the following question: 'If God did have a reason for allowing suffering, would I likely know about it?' Skeptical theists want to say 'Hell no! The cognitive gap between you and the Omni-God is vastly greater than the gap between an ant and a human being. If the Omni-God does exist, then you should not expect to understand God's reasons for acting as She does.'

The skeptical theist's response to the problem of evil is ingenious. And if we set up the problem of evil as a matter of how we get from 'we can't think of why the Omni-God would allow evil' to 'there probably is no reason that would justify the Omni-God allowing evil,' then I think the skeptical theist's way of defusing the problem of evil might be effective. To be fair, this is the way the problem of evil has been construed in recent decades, since an influential version of it was put forth by William Rowe in the 1970s.[19]

However, my approach to the problem of evil is a little different. My argument does not involve an inference from 'we can't think of why the

Omni-God would allow evil' to 'there probably is no reason that would justify the Omni-God allowing evil.' Rather, I argue directly for what I called above the 'Cosmic Sin Intuition':

The Cosmic Sin Intuition: It would be immoral for an all-powerful being to deliberately create a universe like ours.

How confident can we be about the Cosmic Sin Intuition? From my admittedly limited perspective, the Cosmic Sin Intuition seems correct. But skeptical theists question how we can know that the goods and evils we are aware of are representative of all the goods and evils that bear on our situation. For a pollster to get trustworthy results, they need to ensure the group of people they have sampled is properly representative of the electorate as a whole, in terms of age, race, income, etc. But if the Omni-God has knowledge of a much greater range of goods and evils than we do, then it could be that from the Omni-God's perspective—taking into account this much greater range of goods and evils—our universe ends up being the best possible world to create. In other words, maybe from God's perspective, the Cosmic Sin Intuition is obviously false.[20]

I concede that we can't know for sure that the goods and evils we know about are representative of all the good and evils that bear on our situation. But nor do we have any reason to think they aren't. At the end of the day, all we can do is work with the evidence we have and judge as well as we are able. Physicists now think dark matter makes up 85 per cent of our universe, making the matter we are able to observe vastly unrepresentative of all the matter there is. It could well be that if we could observe dark matter, this would radically undermine our current best theories of more familiar kinds of matter, by providing evidence for some totally new Grand Unified Theory of both kinds of matter. Does that mean our current theories are totally unsupported? Of course not. The rational imperative is to believe the theory best supported by the evidence we do have.

Similarly, I cannot rule out that there might be all sorts of goods and evils I have no clue about, and maybe if I took them into account, the Cosmic Sin Intuition would lose its force. But we do have substantial knowledge of morality, and no reason to think that this knowledge isn't up to the task of assessing the Cosmic Sin Intuition. Skeptical theists might argue that the possibility of theism—or a moral agent so much greater than any of us—should undermine our confidence in our moral judgements. But nobody argues that this possibility undermines other moral judgements that we are confident of, such as the intuition that slavery is wrong. Why should matters be different when it comes to the Cosmic Sin Intuition? As is always the case, we just do the best we can with the evidence we have. Recall, the position of the skeptical theist is that the evil and suffering we find in our world provides *no evidence whatsoever* against the existence of the Omni-God, and hence that attempts to determine whether or not God exists by assessing the Cosmic Sin Intuition through careful moral reflection, trying to construct theodicies, and the like, are no better than coin tossing. But suppose your life depended on working out whether or not the Cosmic Sin Intuition was true. Would you try to make a reasoned judgement based on humanity's collective moral knowledge, or would you toss a coin? I know which I'd do.[21]

Moreover, if we really are too ignorant of morality to judge what an Omni-God might have good reason to do, then we can't rule out that an Omni-God might have good reason to create a universe in which everybody has unremitting pain for millions of generations, and hence that if we found ourselves in such a world, we'd have no reason to doubt it had been made by a loving creator. This reminds me of the ghost of the monk in Philip Pullman's *The Amber Spyglass* who, having experienced the bleak land of the dead where all—saints and sinners—end up, nonetheless refuses to stop believing in God's promise of a blissful afterlife:

> The world we lived in was a vale of corruption and tears. Nothing there could satisfy us. But the Almighty has granted us this blessed place for all eternity, this paradise, which to the fallen soul seems bleak and barren, but which the eye of faith sees as it is, overflowing with milk and honey and resounding with the sweet hymns of angels. *This* is heaven, truly!

It's not a perfect analogy. A better one would be if the ghost monk was clear-sighted about the misery of the land of the dead, but, after a million years of living there, still clung to the belief that his loving creator had put him there for a good reason. Clearly, this would be irrational, and in these circumstances the only rational conclusion to draw would be that the Omni-God does not exist. Our circumstances are not as extreme, and so we should not be *as* confident in the non-existence of the Omni-God as we would be after a million years of universal, underserved, and unrelenting misery. But there is no difference here in principle. Observing the tragedies of the world, we also ought to conclude that the Omni-God probably doesn't exist.

I am not certain that the Cosmic Sin Intuition is sound. I cannot rule out with 100 per cent certainty the possibility that there is an all-powerful and loving creator who had some very good reason for allowing the terrible suffering we find in the world. But the Cosmic Sin Intuition seems as solid as any other moral intuition. Moreover, it has been rigorously tested by the attempts of our best theologians to construct possible reasons God might have for allowing suffering. Due to space constraints, I've only considered one—very influential—theodicy, and readers who want a more informed opinion should explore other attempts to reconcile God with the reality of suffering. My own considered opinion is that Plantinga is right: they all come up short. Unlike Plantinga, however, I infer from this that we have very good reason to accept the Cosmic Sin Intuition. And the Cosmic Sin Intuition entails that the Omni-God does not exist.

Logic or Evidence?

I want to finish by discussing how the case I have built connects to a distinction which will be familiar to anyone who has studied this topic, namely the distinction between *logical* and *evidential* versions of the problem of evil. The standard line in the philosophical literature goes as follows:

- *In the old days*, atheist philosophers tried to demonstrate that God's existence is *logically incompatible* with the existence of evil and suffering, and hence that, given evil and suffering exist, God's non-existence can be logically demonstrated. This approach is known as the *logical* version of the argument from evil.[22]
- *Nowadays*, most philosophers on both sides of the debate accept that the logical version of the argument is too strong. Even if we can't think of any good reasons why God might have allowed suffering, we can't *logically demonstrate* that there are no such reasons. Thus, contemporary atheists tend to argue not that evil and suffering *logically entail* God's non-existence, but rather than they are *good evidence* for God's non-existence.

Where does my argument fit into this dichotomy? You might think that I am making a version of the logical argument from evil, given that God's non-existence can be logically demonstrated on the basis of the Cosmic Sin Intuition:

The Cosmic Sin Intuition: It would be immoral for an all-powerful being to deliberately create a universe like ours.

Therefore, if our universe has a creator, She is either not all-powerful or not perfectly moral (or both).

Therefore, there is no such thing as the Omni-God.

However, note that this is not a logical inconsistency between **the Omni-God** and **the evil and suffering we find in the world**, but rather between three claims:

(i)   the Omni-God exists,
(ii) we are not hallucinating or imagining the evil and suffering we seem to find in the world,
(iii) the Cosmic Sin Intuition is true.

In other words, it is logically impossible for all of (i)–(iii) to all be true.

Strictly speaking, then, I am not giving a version of the logical argument from evil, because I'm not trying to demonstrate an incompatibility between the Omni-God and evil and suffering. It is commonly assumed that the only alternative to the logical approach is to argue via some kind of inductive inference that God's existence is highly unlikely. For example, the entry on the 'Problem of Evil' in the *Stanford Encyclopaedia of Philosophy* states:

> If [God's lack of reasons for allowing the suffering we find] cannot, at least at present, be established deductively, then the only possibility, it would seem, is to offer some sort of inductive argument in support of the relevant premise. But if this is right, then it is surely best to get that crucial inductive step out into the open.

An inductive inference is one in which we infer from something that is true of a sample of a certain group of things to something that is true of the group as a whole. At the time of writing, the pollster YouGov has just reported that 54 per cent of the UK would vote for the Labour Party if a general election were held immediately. Now, of course, YouGov didn't interview every single person in the country. Rather, their approach is that they find out what a sample of the population thinks, and then try to infer from that something that is probably true of the population as a whole. Whether this inductive inference is justified will depend on whether the sample is representative, in terms of age, gender, socioeconomic status, etc. Science is highly dependent on inductive inferences. We have not observed everywhere in the universe, and yet scientists take it to be reasonable to infer that the laws of nature that hold everywhere we have observed probably also hold in parts of the universe which we haven't yet observed.

In terms of the argument from evil, the inductive inference in question would be something like the following:

- In terms of the moral considerations we know about, an Omni-God would not have good reason for allowing the suffering we observe.
- Therefore, probably in terms of all moral considerations, an Omni-God would not have good reason for allowing the suffering we observe.

This way of setting up the argument from evil falls prey to the skeptical theist's critique. Pollsters have ways of working out whether or not their sample is representative of the whole population. But how do we know the moral considerations we know about are suitably representative of *all* moral considerations. If this were the only alternative to the logical argument from evil, the skeptical theist response would be strong.

However, an inductive inference is not the only alternative to a logical deduction. We can rather argue directly for a certain moral claim, in precisely the way we standardly argue for moral claims. When arguing that slavery is morally wrong, I don't make a case that slavery is wrong *in terms of all the moral considerations I know about* and then inductively infer that slavery is wrong *in terms of all moral considerations*. Rather, I simply build a case that slavery is wrong, which is of course going to be in terms of the moral considerations I know about (what else?). I then take myself to be justified in believing that slavery is wrong, given I've done the best I can to evaluate this moral proposition. I can't know this for certain; *maybe* there are some moral considerations I don't know about which would make a difference. But certainty is too high a standard. None of us has any qualms about feeling highly justified in believing that slavery is wrong. Why should the matter be any different when it comes to the Cosmic Sin Intuition?

I agree with the common view of philosophers and theologians that the old school form of the logical problem of evil is too ambitious. But I disagree with the consensus view that the only alternative way of running the argument from evil and suffering is via some kind of inductive inference. By defending the inherent plausibility of the Cosmic Sin Intuition, we can build a very strong—although not logically certain—case against the existence of the Omni-God, one which avoids the skeptical theist critique.

The Omni-God probably doesn't exist. Time to look for alternatives!

5

Cosmic Purpose without God

People get stuck in dichotomies of thought. If you don't like Communism, you must be in favour of US-style capitalism, right? Well, not if there are political options other than those two (which of course there are). Another dichotomy is that between the traditional God of Western religion and materialist atheism. If you don't believe in an all-powerful, loving creator, then you must think we live in a meaningless universe, right? Whose team are you on, Dawkins' or the Pope's??

Western philosophy has massively neglected options between these two extremes.[1] In this chapter and the next we will explore three possibilities:

- Non-Standard Designers: Intelligent cosmic designers, but without the perfect qualities of the Omni-God.
- Teleological Laws: Impersonal laws of nature with goals built into them.
- Cosmopsychism: The idea that the universe is a conscious mind with purposes of its own.

Once we have a grip on what might underlie cosmic purpose, we will spend Chapter 7 exploring the implications for human existence.

Non-Standard Designers

Perhaps the most intuitive way of accounting for cosmic purpose is in terms of the intentions of a cosmic designer. In Chapter 4 we ruled out the Omni-God; but the Omni-God is only one kind of cosmic designer. Are there other cosmic designer hypotheses that can account for both the evidence of cosmic purpose explored in Chapters 2 and 3 and the reality of evil and suffering discussed in Chapter 4? Let's explore some options.

The Evil Designer Hypothesis

If the universe is designed, it's designed in such a way that any lifeforms that emerge are destined to suffer terribly, whether that be from natural disasters, the torturous process of natural selection necessary to bring about intelligent life, or just good old-fashioned ageing and death. One might think that the simplest way of making sense of both of these data points—cosmic purpose and horrible suffering—is to postulate an evil creator. Perhaps the designer of our universe delights in suffering and misery, and created the world in order to subject her creation to the horrors of pain and loss.

It's a nice idea, but, as Stephen Law has pointed out, the Evil Designer Hypothesis faces a 'problem of good': the mirror image of the problem of evil facing the Good Designer Hypothesis.[2] If the Evil Designer wants to create misery, why did she create all the good things? Why create love and beauty, the smile of a baby, and the song of the nightingale?

Law imagines 'anti-theologians'—defenders of the Evil Designer Hypothesis—cooking up 'anti-theodicies,' i.e. explanations of why the Evil Designer allows good things to exist, to mirror the theodicies offered in support of the Good Designer Hypothesis. Maybe there could be a free will anti-theodicy: the Evil Designer doesn't just want us to be robots programmed to cut each other up, she wants us to *freely choose* to do bad things. After all, only freely chosen actions can count as genuinely evil. To allow for that possibility, the Evil Designer needs to give us free will. Of course, if you give people free will, then unfortunately they're sometimes going to freely choose to smile and help old ladies across the road, but that's just part of the deal. Or maybe anti-theologians would press a 'greater evils' justification: the Evil Designer creates good things for the sake of some greater evil. It's only when you've experienced love that you can really feel loss; it's only when you've tasted happiness that the pain of misery is keenly felt. Ultimately, Law argues that these anti-theodicies designed to save the Evil Designer Hypothesis would be just as implausible as real-world theodicies formulated in defence of the Good Designer Hypothesis, and hence the Evil Designer Hypothesis is just as implausible as the Good Designer Hypothesis. That seems about right to me.[3]

But perhaps it's wrong to think of our creator as a kind of cartoon villain intent on causing mayhem and suffering. Maybe the cosmic designer is a kind of Hannibal Lecter character, who loves beauty but has no empathy, and has created this universe because it's a beautiful, wonderful thing, despite the great suffering it contains. Paul Draper calls this hypothesis 'aesthetic deism.'⁴ Unfortunately, aesthetic deism faces a 'problem of the ugly' which on the face of it is just as intractable as either the problem of evil for the Good Designer or the problem of good for the Evil Designer. As Clayton Littlejohn, a professor of philosophy at King's College London, put it to me on Twitter, 'You've seen Luton airport, right?' Why would an all-powerful being motivated by beauty allow humans to create such monstrosities as Luton airport?!

The Simulation Hypothesis

The Evil Designer Hypothesis departs from traditional theism as regards the moral character of our creator. However, it is implicitly assumed in such discussions that, like the Omni-God, the Evil Designer is in all other respects indiscernible from the Omni-God. In particular, like the Omni-God, the Evil Designer is envisaged to be a non-physical transcendent being, and the ultimate cause of everything that exists.

But why suppose the designer of our universe has such a grand status? Maybe she's just an ordinary Jo in the next universe up. One such possibility is raised by one of the most intriguing philosophical thought experiments of recent times: the simulation hypothesis.

Technology is advancing at a rapid rate. With ever greater computing power, we are able to create increasingly complex simulations of the reality around us, from weather systems to economic modelling to the first moments of the universe. If this continues unchecked, then at some point in the future it may be possible to create a perfect simulation of a human brain, a society of people, perhaps even a small universe. Contemplating the possibility of simulated brains and people prompts a fascinating philosophical question: Would a simulation of you be conscious? Would it have your thoughts and feelings?

An initial response might be: obviously not. A simulation of a hurricane doesn't make the computer it's running on wet. It seems similarly absurd to think that a mere simulation of your brain would actually produce your consciousness. On the other hand, if it were a perfect simulation, then the simulation of your brain would have all of the structure of your actual brain. Perhaps this would be enough for consciousness?

The crucial question here is whether consciousness is 'substrate independent.' That is to say, is it merely *structure* that's important for consciousness—how parts of a system are arranged and the causal connections between them—or does consciousness require a particular kind of *stuff*? Do we need warm and wet flesh and blood to get consciousness? Substrate independence is the thesis that structure is all that matters for consciousness.

To focus the issue, imagine a silicon robot indiscernible from a human being in terms of its behaviour and the information processing in its brain. If consciousness is substrate independent, then the robot would be conscious, as its parts are arranged in the right way.[5] But if consciousness requires wet, biological flesh and blood, then the silicon robot would not really feel anything, despite the impression to the contrary its behaviour will no doubt give.[6]

One reason we might be interested in the question of substrate independence is that the answer will determine whether or not we might one day be able to achieve immortality by uploading our minds onto the internet. If having the right structure is the only important thing for consciousness, then so long as the relevant structure is preserved in the upload, my conscious mind will live on online, staying in touch with my grandkids (and potentially their grandkids and their grandkids, and so on…) via email. However, if the biological *stuff* of the brain is what is required, then the upload will not preserve my consciousness. We can imagine a dystopian scenario in which the entire human race gleefully uploads, thinking it's transitioning to a Utopian afterlife when in actual fact it's unthinkingly committing mass suicide.

Let's assume, for the sake of argument, that consciousness is substrate independent. It follows that a perfect computer simulation of you, because it preserves all of the structure and functioning in your body and brain, would actually have all of your feelings and experiments. What it's like to be a simulation of you would be indiscernible from what it's like to be you. This starts to raise deep philosophical questions, most notably: How do you know you yourself are not in a simulation?

So far, all we have is a skeptical worry, of the 'How would you know?' variety. However, in the early 2000s, Swedish philosopher Nick Bostrom formulated a compelling argument to the conclusion not only that it's *possible* that we live in a simulation but that it's *probable* (given certain plausible assumptions).[7] The thought is that, barring the (not entirely far-fetched) possibility that all civilizations destroy themselves relatively early in technological advancement, it is likely that at least one civilization will reach the point where they have enormous computing power available to them. One thing such advanced civilizations may be interested in doing with this immense computer power is creating 'ancestor simulations,' by which

Bostrom means simulations of earlier periods of their history. With such incredibly powerful computers, they would be able to run many, many such simulations. Unless our descendants are totally uninterested in creating ancestor simulations, it seems likely that at some point simulated human communities will come to outnumber actual flesh and blood human communities. On this basis, there seems a reasonable probability that in the fullness of time, there will be many more simulated human societies than flesh and blood human societies. Now consider, as if from a God's-eye point of view, the broad stretch of human existence: past, present, and future. What are the odds that we happen to be in the presumably quite small minority of human societies that are not running inside a computer? Not great, you might think.

Regardless of how convinced you are by Bostrom's 'simulation argument,' the simulation hypothesis should be taken seriously as a possible explanation of cosmological fine-tuning. Perhaps our creator is just a normal scientist in a technologically advanced civilization where simulated universes can be created with ease. To avoid both the problem of evil for the Good Designer Hypothesis and the problem of good for the Evil Designer Hypothesis, we can suppose that our creator has some purpose independent of how well or badly humans and other animals are doing. Perhaps our creator just wants to study how evolved civilizations will interact. To do this, she needs to simulate a universe where life is possible, so she ensures that the constants in her simulation will be fine-tuned. This is a hypothesis in which our universe has a purpose, but not the kind of purpose religion typically envisages. We exist to serve the intellectual advancement of our creator.

This explanation of fine-tuning perhaps may face a version of the problem of evil, analogous to the problem of evil for the Omni-God we discussed in Chapter 4, depending on how likely you think it is that technologically advanced societies will have some minimal moral standards. Indeed, we suggested in Chapter 4 that a future society in which simulations are possible might set up a 'Committee for the Assessment of the Moral Permissibility of Simulated Worlds.' So long as the members of the committee want to avoid senseless suffering, it is unlikely that they would agree to allow someone to simulate a universe like our own, casting doubt on the hypothesis that we live in a simulation. On the other hand, this objection is much weaker than the problem of evil for the Omni-God, given that the Omni-God is by definition perfectly good whereas there is considerable doubt as to the moral credentials of our simulators.

My biggest concern with the simulation argument is the thesis of sub-strate independence itself. I think my consciousness depends on the specific *stuff* contained within my brain, and thus a simulation of my brain would not be conscious because it's running on a computer made out of different stuff. However, discussion of this topic gets a little bit technical, so I've saved it for the 'Digging Deeper' section of this chapter.

The simulation hypothesis is not the only way of making sense of a non-supernatural designer of our universe. In the 'Digging Deeper' section of Chapter 2, we discussed *cosmological inflation*, the hypothesis that our universe began with a brief period of very rapid growth. The theoretical physicist who came up with this theory, Alan Guth, has suggested that inflation may allow for the possibility that a technologically advanced society could create a universe.[8] If it weren't for inflation, one might assume that naturally evolved creatures wouldn't have the resources to create all of the matter required to fill a whole universe. However, the process of inflation is an explosion of creativity which generates new matter at superfast speed. If future scientists could artificially kickstart inflation, they may be able to give birth to a new universe. Wouldn't the emergence of this fledgling universe annihilate the lab it was created in? Guth proposes that quantum tunnelling may allow a disconnected space to be opened up inside the spacetime continuum of the creators, within which the new universe could evolve.

Contemplating this possibility invites the speculation that our own universe may have been created in this way by some super-advanced civilization. It's an intriguing possibility. Unfortunately, it doesn't help us to explain fine-tuning. According to the physics of inflation, at least as standardly understood, the universe of our creators would also have to be fine-tuned to be compatible with life. Thus, even if our creators fine-tuned our universe, we would be left with the puzzle of who or what fine-tuned *their* universe. The explanation of fine-tuning is merely deferred. Many proponents of inflation, Guth included, hope to explain the fine-tuning via the postulation of a multiverse. I argued in Chapter 2, however, that the multiverse fails as an explanation of fine-tuning.

We could, of course, posit that the universe of our creators has some totally unknown physics. But at this stage the non-supernatural Designer hypothesis starts to lose any motivation. The attraction of Guth's speculation of our origins is that it's merely postulating more of the spacetime continuum we already believe in. What we are now contemplating is an entirely new spacetime continuum with radically different—and completely unknown—physics. It's unclear what advantage this would have over the much simpler postulation of a single supernatural designer.

An Amoral Designer

We have found problems with a good designer and a bad designer. What about an amoral designer? Maybe the cosmic designer has no conception of good or bad, but just has a basic desire to create a universe resembling the one we find ourselves in?[9]

Without saying more, there is a worry the design hypothesis is now starting to look vacuous. If there is no constraint on the kinds of desires we could ascribe to a cosmic designer, then the cosmic designer hypothesis ends up entirely lacking predictive power. However the universe turns out, we could just say it was created by a designer who wanted to make a universe *like that*. This kind of very broad design hypothesis fits equally well with any evidence, and hence explains nothing.

The way around this problem is to limit ourselves to positing designers who are responsive in some way to facts about objective value. As we discussed in Chapter 2, what is so striking about the fine-tuning is that the fixed numbers in our physics fall in the narrow range compatible with a universe containing things of great value, a universe in which there is intelligent life as opposed to just hydrogen. It is this startling fact that justifies taking seriously the possibility of a cosmic designer who cares about value. Whilst a hypothesis that the fixed numbers in our physics were determined by *a designer* is too general to predict anything, the hypothesis that the fixed numbers in our physics were determined by *a designer who cares about value* predicts that the numbers must be compatible with the emergence of a universe containing things of value. This is why the postulation of a value-responsive designer can potentially explain the data of fine-tuning.

It is highly controversial among philosophers whether there are objective facts about value (something we explored in the 'Digging Deeper' section of Chapter 1). I used to think that the fine-tuning argument for a value-responsive designer had force only if we already believe in objective value. However, I now think we can simply build the reality of objective value into the hypothesis we are arguing for on the basis of fine-tuning. In other words, our hypothesis is not simply *that a designer determined the fixed numbers in our physics* but that *there are objective facts about value and a designer was responding to those facts in determining the fixed numbers in our physics*. This way of explaining fine-tuning delivers not simply cosmic purpose but also the reality of objective value (thus saving us from the horror of value nihilism we worried about in the 'Digging Deeper' section of Chapter 1).[10]

A value-responsive designer is not the same thing as a perfectly good designer.[11] Maybe the designer is a mix of good and bad, often creating value but sometimes getting cross and lashing out. Or maybe there are two designers, one good and one bad, as the Manicheans believed, the religion of Saint Augustine before he converted to Christianity. It's not obvious, however, that such stories could explain the precise forms of good and bad we find in our universe, at least in any simple and elegant way. What exactly are the moral failings of the designer that resulted in her creating a universe that will produce life through the torturous process of natural selection rather than a universe from which life will emerge quickly and painlessly? How are powers over creation shared out between good and bad designers in such a way as to lead to a universe that is both fine-tuned for life but also contains very little life? I'm open-minded on this option, but I can't help thinking that filling in the details here is going to lead to quite complex and inelegant hypotheses.

Moreover, there is a much simpler way of tailoring a cosmic design hypothesis to account for the mix of good and bad we find in our universe, which leads us to my personal favourite non-standard designer hypothesis.

The Limited Designer Hypothesis

To my mind, the most plausible way of getting a non-standard designer is by tweaking not the moral character of the Omni-God but rather her power. The 17th-century philosopher Leibniz said that we live in the best of all possible worlds. I don't think that's true: a universe with less suffering would be better. But perhaps we live in the best universe our creator was able to bring into existence. Maybe the designer of our universe wanted to create intelligent life, but simply isn't able to create complex physical systems out of nothing, or by breathing spirit into the dust. Maybe our creator is only able to create from a very simple starting point, such as a Big Bang singularity, and has limited flexibility in the kinds of laws of nature she can establish in her universes; she can fiddle with the numbers, but that's it. Hence, the only way she can create intelligent life is by creating a universe with the right numbers, so that life will eventually evolve. This cosmic designer knew that this would create a hell of a lot of suffering along the way, and was pained by this fact, but it was either that or nothing, and she judged, somewhat reasonably, that it would be better to have the imperfect universe we find ourselves in rather than no universe at all.

Whenever I raise this possibility (did I mention I spent too much time arguing on Twitter?), people always ask what it is that is supposed to be limiting the designer's powers. Usually, this point is raised as an objection: if the cosmic designer is the fundamental entity from which all being derives, then, by definition, there couldn't be anything limiting her. But I don't see why we have to think that something outside of the cosmic designer is limiting her in order to posit limitation to her powers. Explanations have to end somewhere. Why can't it just be a fundamental fact about the cosmic designer that she is able to do some things but not others? Why think that a cosmic designer has to be all-powerful?

I've also heard the objection that a limited cosmic designer would herself be fine-tuned: just powerful enough to fine-tune the universe, but not powerful enough to stop us suffering. But this implies that there's something special and requiring explanation about there being just this level of value/disvalue in the universe, and that's at least not obvious. The fine-tuning data, in contrast, does need explaining, as the numbers of our physics are in the narrow range that allows for the possibility of a universe containing things of great value. If the Limited Designer Hypothesis were very inelegant and non-simple, then I think there would be a criticism here. But the constraint on powers under consideration here is pretty simple: the cosmic designer can only create from a Big Bang singularity a universe with physics of a particular form. This is a simple hypothesis that accounts for the data as we find it. What's not to like?

It is also possible that there is a limited designer that is not quite so limited as to be tied to the form of physics we find in our universe, but who chose a form of physics in which the life-sustaining constants are fairly rare in order to leave evidence of cosmic purpose. Making sense of this would require formulating a hypothesis according to which the creator has a slightly wider range of forms of physics available to her, including ones which would not need to exhibit fine-tuning in order to sustain life, but where there'd be no obvious advantage to choosing a form that doesn't require fine-tuning. It's above my paygrade to work this out, but if such a hypothesis did lead to a gain in simplicity, this might give us reason to think that a limited designer, unable to make her presence known in any other way, chose to leave her signature on the universe in the form of fine-tuning.

In summary, whilst I reject the Omni-God hypothesis, I am open to the possibility of a non-standard design hypothesis. However, design is not the only way to make sense of cosmic purpose, as we will discover in the next section.

Teleological Laws

Thomas Nagel is one of the most influential philosophers of recent times. The way in which Anglophone philosophy defines a person's conscious experience—as 'what it's like' to be that person—is due to Nagel.[12] He has also written ground-breaking political philosophy, on themes we will discuss in Chapter 7.[13] In 2012, however, Nagel wrote a book which caused scandal among his peers.[14] Nagel had reached a point where he had become persuaded that certain features of human beings—consciousness, reason, moral awareness—could not be accounted for in our current scientific worldview of neo-Darwinian materialism. No doubt if he had been arguing for the existence of the Omni-God, he would have received polite reviews from religious philosophers and been ignored by everyone else. As it was, Nagel's book *Mind and Cosmos* rejected the God option as well. The result was a series of very aggressive and unreasonable reviews, rejecting *Mind and Cosmos* as 'an instrument of mischief'[15] and the 'shoddy reasoning of a once great thinker.'[16] Nagel had committed the heresy of not fitting into acceptable categories of thought and was not going to be allowed to get away with it.

What Nagel had realized is that there is no incoherence in the idea of cosmic purpose without God, provided we can expand our conception of the laws that govern our universe. The laws of nature we have been used to for the past 500 years move *from past to future*—ensuring that what happens at earlier times determines what happens at later times; Newton's law of gravity, for example, determined the force any two objects will exert on each other dependent on their masses and the distances between them. Nagel's proposal is that there may also be laws that move *from future to past*—ensuring that the present is shaped by the need to get closer to certain goals in the future, such as the emergence of life. In other words, there may be laws of nature with *goals* built into them. We call these 'teleological' laws, from the Greek word 'telos,' meaning *purpose* or *goal*.[17]

It's a little bit hard to make sense of the idea of purposes existing in the absence of any kind of mind or consciousness. But we must recall that the concept of a 'law of nature' itself was initially connected with the idea of a divine lawgiver, and we now have no problem with the idea that there can be laws of nature in the absence of God. Perhaps we can similarly get used to the idea that cosmic purpose might exist without a designer whose purpose it is.

How do teleological laws connect to more familiar non-teleological laws, such as those specified by general relativity or quantum mechanics? Nagel builds on a highly detailed conception of teleological laws formulated by John Hawthorne and Daniel Nolan.[18] In Chapter 1, we briefly discussed the 19th-century philosopher Samuel Alexander's idea that there was an impersonal force—*Nisus*—that was driving the universe towards higher states of being. Hawthorne and Nolan provide logical precision to this vague idea.

Without getting into all of the complexities, the basic idea is that the teleological laws fill in the gaps that arise from indeterminacy in the laws of physics. To take a concrete example, suppose the following:

- At a certain moment in the history of the universe, the laws of physics leave it open whether X or Y will happen at the next moment.
- There is a teleological law that directs the universe towards the emergence of life.
- If X happened rather than Y, the universe would be brought closer to the emergence of life.

In these circumstances, the teleological law directing the universe towards life will kick in, ensuring that X rather than Y occurs, and thus the universe gets a little bit closer to the goal of life emerging. Over long periods of time, the universe is slowly edged towards that goal.

Nagel does not specify what indeterminacies in the laws of physics might allow room for the teleological laws to do their work. On many interpretations of quantum mechanics—although not the pilot wave theory we explored in Chapter 3—the laws do indeed leave matters indeterminate. However, Nagel's theory would almost certainly be falsified already if the teleological laws kicked in whenever there is *any* kind of quantum indeterminacy. This is because, at any moment, there is such an enormous range of possible happenings which quantum mechanics tells us have *some* chance of occurring, albeit a very low chance. For example, quantum mechanics tells us that there is some incredibly small chance that in the next moment my body will pass through the solid chair I'm currently sitting on, leaving me lying on the floor beneath it. With this amount of flexibility to determine what will happen—choosing, as it were, from this incredibly wide range of options—the teleological laws would be close to omnipotent. Life would be abundant in the universe.

We can avoid this problem if we suppose that the relevant indeterminacy existed only in the very early history of the universe. Indeed, we can sneakily steal some ideas from multiverse theorists to help us here. As we explored in the 'Digging Deeper' section of Chapter 2, some scientists and philosophers try to explain fine-tuning by postulating a huge number of universes, each with slightly different numbers in their physics. Multiverse theorists build on string theory in proposing that in the first split second of the Big Bang, it was indeterminate what values the constants of our universe would have—the strength of gravity, the weight of electrons, etc.—and that random processes determined what they would be. The multiverse theorist hopes that a commitment to many universes will allow us to avoid cosmic purpose, the thought being that if there are enough universes, then it is not so implausible that random processes might have fluked upon fine-tuning in our case (if you have an infinite number of monkeys banging away on typewriters, one of them is eventually going to write Hamlet).

I argued in Chapter 2 that this attempt at explaining the fine-tuning in terms of many universes is fallacious. However, a proponent of teleological laws may agree with the multiverse theorist that at some point in the past it was indeterminate what values the constants of our universe would have, whilst explaining the fine-tuning of our universe in terms of teleological laws rather than many universes. We need not locate this moment before the existence of our universe. In the first 10^{-43} seconds of our universe, known as the Planck Epoch, our current physical theories break down. Perhaps, as many string theorists suppose, the currently unknown laws of physics governing in this period left it open what values our constants would have, leaving room for the teleological laws to do their work. Recall that, when the laws of physics leave it open which option from a range of possibilities will transpire, a teleological law aimed at life will ensure that the possibility most conducive to life will be actualized. In our envisaged scenario, the possibility most conducive to life was the universe emerging from the Planck Epoch with fine-tuned constants.

Can teleological laws provide a satisfying explanation of fine-tuning? Even accepting that the hypothesis of cosmic purpose without God makes sense, there is something problematic about ending our explanation with a teleological law directing the universe towards something which happens to be of great value, such as life. Presumably the teleological laws could have pointed to anything. We might have found ourselves in a universe directed towards non-life, or towards ceaseless suffering, or towards something totally pointless, like having all the matter in the universe arranged in a polka-dotted fashion. It seems, therefore, rather fortuitous that, of all the goals our universe might have had, it happens to be directed towards something of great value. Does this push us back to the idea of a *good* designer, who instituted the teleological laws in order to ensure that the universe is directed towards something of value?

Maybe. Another option is to accept that the universe has the goals it does *because they are good*, but to insist that there is no deeper explanation of why the universe has good goals. John Leslie has defended something like this view, calling it *axiarchism*.[19] As I have said a number of times in this book already, explanations have to end somewhere. An atheist may take it to be brute fact that the constants of physics are fine-tuned for life. A believer in teleological laws explains the fine-tuning in terms of a teleological law pointing to life. A proponent of axiarchism just takes the explanation one level deeper—the teleological law points towards life because life is a good thing—but then stops.[20] In each case, there is a good reason to move to a deeper explanation: (A) taking the fine-tuning of the constants as unexplained is problematic because it implies that, of all the values these constants might have had, they just happened by chance to have the right values for life, (B) taking the teleological laws as unexplained is problematic because it implies that, of all the goals these laws might have aimed at, they just happened to aim at goals that greatly increased the value of the universe.

As we discussed in Chapter 4, Aristotle thought that organisms had essentially goal-directed natures, without feeling the need to postulate a designer who put the goals into things. That view was coherent although arguably disproven by the discovery that biological organisms had been shaped by the blind process of natural selection. However, perhaps the evidence for cosmic purpose outlined in Chapters 2 and 3 gives us grounds for thinking that the universe as a whole is a kind of Aristotelian organism: it exists, in part, to fulfil a good purpose, despite the fact that nobody created it for that purpose. Perhaps the universe is like a plant that, in the fullness of time and with a bit of luck, will grow and blossom into something beautiful.

Teleological laws are the most parsimonious accounts of cosmic purpose. They simply accept the brute existence of cosmic purpose without feeling the need to postulate any deeper explanation of it. On the other hand, the deeper explanation of cosmic purpose provided by non-standard designer hypotheses is an attractive feature. We arguably have a tie here, with one theory ahead in terms of parsimony, the other ahead in terms of explanatory depth. The ideal would be to find a way of securing the extra explanatory depth but with minimal cost in terms of postulating extra entities.

Fortunately, there is such a theory: rather than postulating a supernatural designer, we can instead ascribe mentality to the universe itself. This is the subject of Chapter 6.

Choice Point: There are few different options for what to read next. Chapter 6 explores whether we can account for cosmic purpose in terms of a conscious universe, embedded in a broader discussion of the panpsychist theory of consciousness. In my view, panpsychism offers our best hope for accommodating both consciousness and cosmic purpose, so I hope readers will have a go at this chapter at some point. But if, for now, you've heard enough to be able to make sense of the reality of cosmic purpose, feel free to jump to Chapter 7 where we'll explore the implications of cosmic purpose for human existence. Finally, if you want to stick around for the 'Digging Deeper' section, we will be examining (A) the connection between the hypotheses explored in this chapter and the 'pan-agentialist' view defended in Chapter 3, and (B) the question of whether a simulation of your brain would be conscious (I say 'no').

Digging Deeper

Pan-Agentialism and Cosmic Purpose

To simplify the above discussion, I focused on whether the hypotheses under consideration could account for the data of fine-tuning we explored in Chapter 2. However, there is of course another source of cosmic purpose defended in this book: the pan-agentialist view defended in Chapter 3, according to which matter has a basic disposition to respond rationally to the character of its experience. We turn now to considering how the theories explored above can accommodate pan-agentialism.

Pan-Agentialism and the Limited Designer Hypothesis

I argued above that the Limited Designer Hypothesis is the most simple and elegant form of non-standard designer hypothesis. This theory fits well with pan-agentialism, so long as we specify that the limited designer was able to create rational particles (or more precisely very simple forms of rational energy, as particles did not exist in the first moments of the universe). At the end of Chapter 3, I argued that fine-tuning and rational matter fit together like a key into a lock, and that this is unlikely to be a coincidence. On the Limited Designer Hypothesis, it's not a coincidence. The limited designer created rational matter, perhaps from a range of alternatives she might have created, and placed it in a fine-tuned universe with the aim of rational matter eventually, in the fullness of time, evolving into complex rational organisms. Without the fine-tuning which allows for complex structures, rational matter would have been unable to evolve beyond its most basic state of proto-agency. But fine-tuning without rational matter would have yielded at best a world of zombies or meaning zombies. In the absence of a more direct route to creating intelligent life, the limited designer created a fine-tuned universe with rational matter, and wished it well on its journey.

Teleological Laws

Can teleological laws account for the nice fit between rational matter and fine-tuning, for the way in which they interact to allow for the evolution of experiential understanding? If rational matter has always existed, or has existed since the beginning of time, then presumably teleological laws can't account for its existence. However, the proponent of teleological laws can account for the good fit between rational matter and fine-tuning by holding that it was the great good of allowing rational matter to achieve a higher realization of its nature that the teleological/axiological laws were aiming at when they fine-tuned the constants of physics.

Are Simulations Conscious?

Let's return to the question of whether a computer simulation of my brain would be conscious. As I argued in Chapter 3, I believe we have consciousness not merely because of the way the matter in our bodies and brains is organized, but because:

1. Panpsychism: Our universe is made up of conscious particles.
2. Mental Combination: When arranged in certain structures, those conscious particles combine into conscious systems.

Structure is relevant to the explanation of our consciousness: if the right structure weren't in place, we would not be conscious organisms but mechanisms that happen to be made of conscious particles. But there is no reason to think that the structures that underlie mental combination perfectly correspond to computational structures, and therefore no reason to think that the computational structures realized in a simulation would yield mental combination. Indeed, if the integrated information theory, which I tentatively supported in Chapter 3, turns out to offer the correct account of mental combination, then computers would not generate mental combination. This is because computers—at least anything along the lines of the computers we have today—lack the kind of integration which, according to integrated information theory, is the mark of consciousness.

In his recent defence of the simulation argument, David Chalmers has offered a thought experiment in support of substrate independence.[21] My own thought experiments often concern my imaginary friend Susan; as it happens, Chalmers also picks a Susan as the subject for his thought experiment, although he has real friends to choose from, in this case leading philosopher and AI researcher Susan Schneider. Imagine we slowly uploaded Susan's mind, a few neurons at a time. In a series of operations, successive groups of neurons are removed from her brain, and each time their functioning is taken over by an online simulation of those neurons. Remote signalling devices are installed to ensure that what was previously inputted to and outputted from the physical neurons is now inputted to and outputted from the virtual neurons. Over a few weeks, let us say, Susan's brain is gradually replaced with a virtual brain, one which will remain online after her body has died.

How does this thought experiment support substrate independence? Chalmers' thought is that observing this process happening to Susan would give us the opportunity to test substrate independence, as we could simply ask her how or if her consciousness was affected by the uploading. Now, you might think we'd have to actually *do* the experiment, rather than merely reflect on a thought experiment, to find out what the result would be. But it's not obvious that's true. If the simulated neurons are perfect simulations of the neurons you would have had if you hadn't been uploaded, then, by the very definition of 'perfect simulation,' Susan would behave in exactly the same way as she would if her brain hadn't been uploaded. And therefore, if, at the end of the uploading we ask Susan 'How are you feeling? Has having your brain uploaded made any difference to your consciousness?,' we know—even without doing the experiment—that she will reply 'No, it's all good, thanks.' Assuming we can rely in the normal way on people's testimony about their own conscious experience, we seem to be led to the view that a simulated brain is just as conscious as a physical brain.

Real-life Susan Schneider is not impressed by Chalmers' argument, which is why Chalmers picked her for the thought experiment.[22] Schneider's objection is that this isn't a normal situation, and so we should be more wary than we usually are of relying on people's testimony regarding their conscious experience. The simulation is contrived to ensure that post-upload Susan's body behaves in the same way as it did prior to the upload. And therefore, even if Susan has been transformed into an unfeeling mechanism, this mechanism will still respond as conscious Susan would have done. Perhaps in this kind of situation—artificially set up to produce the behaviour associated with consciousness whether or not the system really is conscious—we cannot trust what the system reports.

Chalmers' counter-response—and this is where the *gradual* aspect of the uploading becomes key—is to demand a plausible hypothesis as to *when* consciousness disappears. It seems rather implausible that the replacement of a few physical neurons with virtual neurons will be what makes Susan's consciousness suddenly disappear. It seems more likely that if Susan's consciousness is going to go, it will gradually fade as more and more neurons are replaced; call this the 'Fading Consciousness Hypothesis.' But, again, so long as the virtual neurons are *perfect* simulations of the physical neurons Susan would have had if it were not for the upload, then, by definition, she will behave just as she would have done without the upload, including what she says in response to questions. It turns out, therefore, that the Fading Consciousness Hypothesis is one in which although Susan's consciousness is slowly fading away, she continues to maintain that nothing is changing! This seems an even more peculiar upshot. A non-conscious mechanism behaving as though it were conscious is one thing, but here we have a person who is conscious but is radically wrong about what their consciousness is like (or at least a person who *says* radically wrong things about what their consciousness is like).

Crucially, Chalmers' argument assumes micro-reductionism, the thesis we explored in Chapter 3 according to which everything an organism does is entirely fixed by the facts of fundamental physics. The simulation essentially replicates how the neurons will behave if their behaviour is entirely determined by the laws of physics. As pan-agentialists—I'm assuming at this point that all readers were totally persuaded by my argument from Chapter 3—we reject this assumption. As there emerge systems with conscious inclinations embedded in a complex web of meaning and understanding, we will start to get deviations from the predictions standardly associated with quantum mechanics. In other words, the emergence—or for that matter disappearance—of experiential understanding *makes a behavioural difference.*

Pan-agentialists, therefore, will reject the claim that Susan will continue to behave the same as her brain is gradually uploaded. We do not accept that Susan's behaviour is entirely determined by the laws of physics, and therefore need not accept that preserving how her parts would behave solely under the guidance of the laws of physics will preserve the behaviour of real-world Susan. If the replacement of physical neurons with virtual neurons does start to make Susan's consciousness fade, she may very well respond to this new situation by saying, 'Gosh, my consciousness is fading…help!'

Of course, once Susan's brain has been entirely replaced by a virtual brain, the behaviour of the virtual neurons will entirely determine her behaviour (at least to the extent that her brain would have determined her behaviour prior to being uploaded). But if the upload has indeed made her consciousness disappear, then, post-upload, micro-reductionism will be true of Susan's virtual brain in a way it wasn't true of her pre-upload physical brain. Whereas the behaviour of Susan's conscious physical brain—the locus of her conscious experience—was determined by Susan's rational responsiveness to her experiential understanding of reality, the behaviour of Susan's virtual brain will be entirely determined by the behaviour of its (virtual) parts. This radical change may very well impair Susan's behavioural functioning. As she loses consciousness, Susan may become less articulate and more mechanical, as she is reduced to a non-conscious mechanism (albeit one composed of conscious particles).

In his argument, Chalmers is essentially assuming micro-reductionism, and then, against the backdrop of that assumption, arguing for substrate independence. If, however, you think you have good reason to *reject* substrate independence, then you can just run Chalmers' argument back to front:

- If micro-reductionism were true, then we could demonstrate (via Chalmers' argument) that substrate independence is true.
- But substrate independence is not true, and therefore cannot be demonstrated to be true.
- Therefore, assuming nothing else is wrong with Chalmers' reasoning, the micro-reductionist assumption on which Chalmers' argument is based must be false.

If it does turn out that micro-reductionism is false, this has the potential to offer us new ways to make progress on the challenges, discussed in Chapter 3, of locating consciousness at the macro level. If micro-reductionism holds, then whether or not a system is conscious will make absolutely no difference to its behaviour, which is totally determined in either case by the arrangements of its particles. But if systems start to behave differently when they gain or lose consciousness, then this will clearly be an important factor in identifying which systems are conscious.

Furthermore, if we accept some kind of libertarian free will (an idea I explored and expressed some sympathy for in Chapter 3, even though it's not an essential commitment of pan-agentialism), not only does Chalmers' argument for substrate independence collapse, but the simulation hypothesis itself is undermined. The simulation hypothesis only makes sense if our behaviour is entirely predictable and hence can be anticipated by an algorithm. If conscious systems with experiential understanding are able to freely respond—in the libertarian sense of 'free'—to the reality that is presented to them, then this will not be the case.

For all of these reasons, I doubt that a simulation of my brain would be conscious, and hence I doubt that the simulation hypothesis can account for fine-tuning. I (consciously) think, therefore I am not in a simulation.

6

A Conscious Universe

Something strange has been going on in academic philosophy of late. A view once laughed at, in so far as it was thought of at all, has entered the mainstream. Like anything in philosophy, it's wildly controversial. But the idea that the reality we live in might, at its fundamental level, be made up of consciousness, is now being taken very seriously indeed. In this chapter, I'll explain why, before exploring how a conscious universe can help us to make sense of cosmic purpose.

What Breathes Fire into the Equations?

Sometimes familiar things can open up new possibilities, when looked at from a different perspective. In the 1920s, philosopher and future Nobel Laureate Bertrand Russell was thinking very hard about the fact that our fundamental science, i.e. physics, is purely mathematical.[1] This is something we take for granted these days, but it was a revolutionary decision of Galileo in the 17th century for 'natural philosophy'—what we now call 'physical science'— to take a purely mathematical form. The mathematics has changed a lot, with the introduction of imaginary numbers, and high-dimensional, non-Euclidean geometries; but, still, physics is basically a bunch of equations.[2]

Of course, if you're a practising physicist, mathematics is very useful, as it allows you to make incredibly precise predictions. But what does it mean for a *philosopher* interested in the fundamental nature of reality that our fundamental science is pure math? The implication is that our fundamental science doesn't really tell us very much about the fundamental nature of reality, it merely describes it in terms of its mathematical structure. As far as physics is concerned, fundamental reality could turn out to be *anything*, so long as it has the right mathematical structure.[3]

In the last decade or so, there has emerged a new generation of Bertrand Russell-inspired consciousness researchers trying to make sense of the idea that *consciousness* underlies the mathematical structures of physics.[4] On one very standard form of this view, what we find at the fundamental level of reality are *networks of very simple conscious entities*. These conscious entities behave in simple, predictable ways, in virtue of their very rudimentary experiences. Through their interactions, they realize certain patterns and mathematical structures. The idea, then, is that those patterns and mathematical structures *just are* what we call 'physics,' and in this way we get physics out of underlying facts about consciousness. When we think about those simple conscious entities in terms of the mathematical structures they realize, we call them 'particles' and we call their properties 'mass,' 'spin,' and 'charge.' But all that really exists at the fundamental level are conscious entities.

For the proponent of 'Russellian panpsychism,' as this view has become known, physics emerges from consciousness. On the final page of *A Brief History of Time*, Stephen Hawking declared that even a final theory of physics wouldn't tell us what 'breathes fire into the equations and makes a universe for them to describe.'[5] For the Russellian panpsychist, it is consciousness that breathes fire into the equations.

If this view is right, it follows that when you're doing physics, you're actually studying the behaviour of very simple conscious entities. This sounds odd. It doesn't feel like this is what you're doing. But that's only because, when you're doing physics, you're just interested in the mathematical structures themselves. You're not interested in what, if anything, underlies those structures. This is a question for philosophers, not physicists.

A few years ago, the physicist Sabine Hossenfelder wrote a blog post critiquing panpsychism.[6] Her basic thought was that there would inevitably be a clash between panpsychism and the standard model of particle physics, our best theory of the twenty-five kinds of fundamental particle. This is because the standard model predicts the behaviour of particles on the basis of their *physical properties*, things like mass, spin, and charge. If particles also have peculiar *non-physical* consciousness properties, then presumably this will impact their behaviour, leading to predictions that differ from the standard model. Given that the standard model is well confirmed, we have good reason to think that it's correct, and therefore that panpsychism is false. To put it more simply: if particles were conscious, then their consciousness would show up in their behaviour, and physicists would spot it in their experiments.

The problem is that Hossenfelder misunderstands panpsychism, or at least the Bertrand Russell-inspired form of panpsychism which has become prominent in contemporary academic philosophy. Hossenfelder is interpreting panpsychism in *dualistic* terms, as though the particle has its physical properties—mass, spin, and charge—and *in addition* certain consciousness properties. That's not the view. Russellian panpsychists are not adding to physics. Instead, they deny that physics is the fundamental level of reality. There is a more fundamental story about consciousness underlying the mathematical structures we get from physics.

Imagine a biologist who reads about electrons and declares, 'What nonsense! Animals behave the way they do because of the things I study: cells, muscles, the pumping of blood, etc. If there were such things as electrons, they would presumably also impact how animals behave, and we biologists would be able to spot them.' The biologist's mistake is assuming that electrons exists at the level of reality he is concerned with. The biologist may have the complete story at the biological level; that is perfectly consistent with the existence of electrons at a more fundamental level. Similarly, Hossenfelder errs in assuming that the micro-level consciousness postulated by the panpsychist exists at the level of reality that is her concern: the most basic mathematical structures. But panpsychist philosophers are concerned with reality at a deeper level: the reality that underlies the mathematical structures identified by physics.

The Mind-Body Problem

Even if panpsychism is *consistent* with what we observe, what reason do we have to take it seriously? My view is that panpsychism offers the best solution to the *mind-body problem*, the philosophical challenge that arises from the fact that we access objective reality in two very different ways: *perception* and *introspection*. In perception we access *the physical world* through our senses, something we've learnt to do more accurately and precisely through science. Through introspection we access *consciousness*, via our immediate awareness of our feelings and experiences. The mind-body problem is the challenge of how to bring these two seemingly very different things—consciousness and the physical world—into a single unified theory of reality.

Broadly speaking, there are three philosophical solutions to the mind-body problem:

1. Materialism: The physical world is fundamental, and consciousness arises from physical processes in the brain.[7]
2. Panpsychism: Consciousness is fundamental, and the physical world arises from more fundamental facts about consciousness.[8]

3. Dualism: Both consciousness and the physical world are distinct but equally fundamental aspects of reality. Contemporary 'naturalistic' dualists like David Chalmers postulate special 'psycho-physical' laws of nature to hook consciousness up to the physical world.[9]

Crucially all three views, at least in certain forms, are *empirically equivalent*, meaning that you can't distinguish between them in any straightforward way with an experiment. This is not surprising, as the mind-body problem is a philosophical rather than a scientific problem. Similarly, you wouldn't expect an experiment to conclusively settle which of the pro-life or pro-choice positions on the ethics of abortion is correct. We must simply try to assess these different options as philosophical hypotheses and try to work out which is more successful on its own terms.

In terms of materialism, nobody has ever made the slightest progress on its central explanatory task of explaining how we can get consciousness out of purely physical processes in the brain. Moreover, I think there are good arguments that show that such a thing cannot be done in principle (I'll present a basic form of this kind of argument in the 'Digging Deeper' section of this chapter). Regarding the central explanatory task of panpsychism, in contrast, we have already shown precisely how it can be done, in terms of the Bertrand Russell-inspired approach described above. Because physics is purely mathematical, so long as fundamental conscious entities, through their interactions, realize the right mathematical structures, we can quite straightforwardly account for the emergence of physics. The problem is solved. The mathematical structures of physics cannot produce consciousness, but consciousness can produce the mathematical structures of physics.

Dualism is a coherent possibility. But as scientists and philosophers we aim to respect Ockham's razor by going for the simplest theory consistent with the data, and this pushes us towards one of the first two theories. Why believe in two kinds of fundamental property when you can make do with one?

I think materialism survives because people think of it as the 'scientific' option. But materialism and panpsychism are *philosophical*, not scientific, rivals. My friend Anil Seth is an eminent neuroscientist of consciousness and an ardent supporter of materialism. We've debated on several occasions, in a fierce but friendly manner, and each time I think we've learnt something from the experience (maybe Anil would disagree...).[10] Anil accepts that neuroscientists haven't yet found a satisfying explanation of how brain processes produce consciousness, but nonetheless argues that the ongoing success of the science of consciousness lends credence to materialism. The problem is that all of the science Anil defends is completely neutral on the philosophical question of what the correct solution to the mind-body problem is. As I read Anil's excellent book *Being You* in preparation for chatting to him on my *Mind Chat* podcast, I realized the only point at which I disagreed was when Anil explicitly stated that he was a materialist, almost as a 'by the way' add-on unconnected to the central ideas in the book.[11]

Consider, for example, Anil's very interesting claim that consciousness is correlated with predictive processing in the brain. The idea is that, whilst the brain is stuck in darkness inside the skull, it is continuously modelling the world around it. As sensory information comes in, the brain checks these models against the information it receives, correcting and updating as and when mismatches arise. Predictive processing is essentially the brain's 'best guesses' as to what's going on, and Anil is sympathetic to the hypothesis that it's this that determines the character of consciousness.

It's a fascinating proposal, with some experimental support. But, like any other proposal about how consciousness is correlated with physical processing in the brain, it fits just as well with panpsychism and dualism as it does with materialism. Each of the theories will simply account for the scientifically established correlation in a different way:

Materialism: Consciousness arises from predictive processing.

Panpsychism: Predictive processing arises from underlying facts about consciousness.

Dualism: The psycho-physical laws connect consciousness to predictive processing in the brain.

Our choice, then, is not between a theory with scientific support and a theory with only philosophical support. Our choice is between a philosophical explanation nobody's ever managed to make sense of (materialism) and a philosophical explanation we know how to make sense of (panpsychism). Once the options are correctly understood as *philosophical*—rather than scientific—rivals, there is, to my mind, an obvious winner.

Is it really so simple? Many think that panpsychism faces a devastating objection, in the form of the *combination problem*. This is the challenge of understanding how many conscious particles come together to form a complex system which has its own unified consciousness. It's no good postulating consciousness at the fundamental level if you can't explain *our consciousness*— the only consciousness we have any pre-theoretical reason to believe in—in terms of it. In a couple of my earliest publications, in the days before my conversion to panpsychism, I tried to argue that the explanatory gap between particle consciousness and brain consciousness is just as serious as the explanatory gap the materialist faces between physical process and consciousness.[12] If the very explanatory gap which panpsychists use to reject materialism reoccurs in the context of panpsychism, it seems we've gotten nowhere.

These days, I think that the combination problem is a concern only for very reductive forms of panpsychism, according to which in some sense really there are only conscious particles, and what we call 'Susan's mind' is really just a complex arrangement of conscious particles. Luke Roelofs' excellent book *Combining Minds* is one of the best attempts to make sense of reductive panpsychism. I'm increasingly skeptical, however, of *any* purely reductive account of human consciousness, whether it's trying to reduce it to physical processes or to conscious particles. For one thing, such purely reductive stories are incompatible with the pan-agentialist theory I defended in Chapter 3, on which conscious organisms are more than the sum of their parts.

If we can't, as the materialist aspires to do, reduce consciousness to something else, doesn't that leave us with dualism? Not necessarily. Even if we can't tell a *fully* reductive story, panpsychists may be able to give a *partial* reduction of the human mind. In my recent paper 'How exactly does panpsychism explain consciousness?', I develop a kind of hybrid of reductive and non-reductive views, by distinguishing between *conscious experiences*, on the one hand, and *the 'I' that has the conscious experiences*, on the other.[13] On this hybrid view, the 'I' is more than the sum of its parts, but its conscious experiences are 'inherited' from streams of consciousness at the level of fundamental physics. This particular model of partial reduction is not available either to materialists or to dualists, given that they don't attribute consciousness at the level of fundamental physics. It may even turn out to be empirically testable whether our conscious experiences are partially reducible, by testing whether the causal properties of the corresponding brain states are partially—but not wholly—reducible to the causal properties of their parts, offering hope of empirical support for panpsychism over its rivals. Having said that, we're a very long way from such experiments being practically feasible.

Even if human consciousness ends up being utterly irreducible, it's still better to be a panpsychist than a dualist. Panpsychism earns its keep through its reduction of the physical world to consciousness. If it can also reduce human consciousness to particle consciousness, whether partially or wholly, then that's a bonus: the more we can reductively explain, the simpler our basic theory of reality. But even if the human mind remains stubbornly irreducible, it's still better to have only one kind of property at the base of reality (consciousness) as the panpsychist does, rather than two (consciousness and physical properties) as the dualist does.

Here's a prediction: in fifty years, further investigation in the brain will reveal the causal dynamics in the brain associated with consciousness to be *strongly emergent*, that is to say not reducible to underlying chemistry and physics. This rules out the materialist view that everything can be ultimately reduced to fundamental physics. At that point, the choice will be between naturalistic dualism and panpsychism (unless the former is ruled out because the partial reductionist picture described above is confirmed).[14] As a result of the greater simplicity and elegance of panpsychism over dualism, panpsychism will, over time, come to seem just obviously correct.[15]

Can Panpsychism Account for Cosmic Purpose?

When people think of panpsychism, they tend to assume the fundamental conscious entities are particles, and the complex consciousness of the human or animal brain is somehow built up from the consciousness of particles (this is how we have thus far conceived of panpsychism in this book). However, many theoretical physicists are inclined to think that the fundamental building blocks of reality are not particles but *universe-wide fields*, and that particles are simply local vibrations within those fields. If we combine a fields-based picture of the universe with panpsychism, we end up with the view that the fundamental forms of consciousness underlie these universe-wide fields, and that a fundamental mind is the bearer of those fields: the universe itself. This hypothesis has become known as 'cosmopsychism,' and I have explored various forms of it in my academic research.[16]

When I tell people about my work on cosmopsychism, they tend to assume it's a kind of pantheism, i.e. the view that the universe is identical with God. However, most cosmopsychists don't think the universe has human-like qualities of *thought*, *understanding*, and *agency*. The kind of experiential understanding enjoyed by human beings is the result of millions of years of evolution, but the consciousness of the universe has not been shaped by the pressures of natural selection. In my first book *Consciousness and Fundamental Reality*, my working hypothesis was that the consciousness of the universe is just some kind of huge meaningless mess. It's going to be very complicated, as the universe has a very complicated physical structure. But the universe doesn't appear to have the kinds of cognitive structure needed for reflective thought.

This 'meaningless mess' form of cosmopsychism ends up being the most plausible form of cosmopsychism if one's primary motivation is accounting for the consciousness of humans and non-human animals, which was my only aim in my first book. However, in this book we are interested in explaining not only consciousness but also cosmological fine-tuning.[17] Working with both of these data-points can lead to a very different picture of the conscious universe.

There are a couple of modifications one has to make to the basic cosmopsychist picture to get you to a hypothesis that can account for fine-tuning. The first is to replace the picture of a universe of messy meaningless experience, blundering from one moment to the next, with a view of the universe as something that recognizes and responds to considerations of value. On the view we can call 'teleological cosmopsychism,' the universe is essentially driven to try to maximize the good.[18]

This might sound like a bold and extravagant claim. However, I believe that David Hume—the 18th-century Scottish philosopher discussed in Chapter 1—was right that we can't directly observe what drives the universe to behave as it does; we can observe *how* the universe behaves but not *why*. Many philosophers postulate impersonal causal powers to explain the behaviour of the universe. But it's equally consistent with observation to suppose that the universe's drive to maximize value is running the show.

Or is it? Doesn't this lead straight back into the problem of evil? If the universe is trying to maximize the good, how do we explain the terrible things that happen within it, at least on the planet we live on? Also, how do we think about the laws of physics on this picture? If the universe is driven by a compulsion to maximize the good, shouldn't there just be one law of physics, one reminiscent of the Scout Law I recited in my youth: 'Do Your Best'?

We can kill both of these birds with one stone. On teleological cosmopsychism, the laws of physics record *the limitations of the universe*. Each moment, the universe is pushing to maximize the good, but under quite severe constraints as to what it is able to do. As with the Limited Designer Hypothesis outlined in Chapter 5, it's not that something outside of the universe is limiting the universe. The universe is the totality of everything there is. It's just a primitive fact about the universe that it is able to do some things but not others.

Even if this view is consistent with observation, why should we take it seriously? The attraction of teleological cosmopsychism is to be found in the explanation it can offer of fine-tuning. In the discussion of teleological laws in Chapter 5, we built upon the speculation of multiverse theorists that in the Planck Epoch—the first split second of the universe when our current models of physics break down—it was indeterminate what values the constants of our physics would have. The teleological cosmopsychist can also adopt this hypothesis, holding that this indeterminacy in the first moments of its existence provided the conscious universe with the freedom to determine the values of its constants moving forward. Rather than postulating a supernatural designer or teleological laws to fine-tune the universe, the teleological cosmopsychist simply proposes that the universe fine-tuned itself.

Even if teleological panpsychism provides *an* explanation of fine-tuning, is it an attractive one? Ockham's razor compels us to keep our theories as *parsimonious* as possible, in the sense of not postulating entities beyond necessity. In fact, contrary to first appearances, teleological cosmopsychism is extremely parsimonious. We know, or so I would argue, that there must be *something* underlying the mathematical structures identified by physics, otherwise our universe would contain no consciousness. And we know there must be *something* that drives the predictable behaviour of the universe.[19] It's certainly possible that the fundamental level of reality is wholly impersonal and non-conscious. But the alternative hypothesis of a universe responding to value under limitations recorded by the laws of physics is empirically indiscernible and no less parsimonious. As we have described the view so far, not a single additional entity has been postulated to explain fine-tuning, beyond those that we already needed to meet other theoretical obligations.

However, there is one more postulation we need to add to teleological cosmopsychism, which does add some cost. If, during the first split second of time, the universe fine-tuned itself in order to allow for the emergence of life billions of years in the future, the universe must in some sense have been aware of this future possibility, in order to act in such a way as to bring it about. To account for this, we can attribute to the universe conscious awareness of the full possible consequences of each of the options available to it. This may sound like a massive cost. But here, again, we can cheekily borrow from multiverse theorists. In defence of their postulation of an enormous number of universes, multiverse theorists argue, quite plausibly in my view, that what's important for Ockham's razor is not so much how many *things* your theory postulates, but how simple the general principles underlying your theory are. Whilst the postulation that the universe has conscious awareness of the full consequences of every choice available to it entails a huge number of conscious experiences, the causal principle governing their generation is very simple. In that sense, this postulation is not a significant cost to the theory. Overall, teleological cosmopsychism is a surprisingly economical explanation of cosmological fine-tuning.[20]

What Would You Say to the Truth Demon?

In Chapters 5 and 6, we have explored three accounts of cosmic purpose:

- non-standard design hypotheses
- teleological laws
- teleological cosmopsychism

I believe that all three of these accounts of the source of cosmic purpose should be taken seriously. But teleological cosmopsychism seems to have the edge. The teleological laws and non-standard designer hypotheses are on a par: the former trumps the latter in terms of parsimony, whereas the latter trumps the former in terms of explanatory depth. Teleological cosmopsychism secures the explanatory depth of the non-standard designer hypotheses but without incurring much of their cost. More bang for your buck! The only real cost of teleological cosmopsychism, over and above what we need to account for consciousness and the regularities we observe in nature, is the axiom that the universe is aware of the total consequences of all possible options available to it. But is this more of a cost than postulating teleological laws? It doesn't seem to me that it is, whereas it provides a more satisfying explanation of fine-tuning than just postulating brute goal-directed laws.

My friend and podcast co-host Keith Frankish likes to pose to other philosophers the 'Truth Demon' thought experiment. If we seriously engage with it, it's a good way of separating oneself from the ego-driven desire to win arguments and really focusing on the question of what's most likely to be true. The thought experiment goes like this:

The Truth-Demon Thought Experiment: The Truth Demon asks for your answer to a certain theoretical question and will send you to hell for eternity if you give the wrong answer.

I'd be pretty terrified of the Truth Demon demanding from me an answer on what the ultimate nature of reality is, as there is a great deal of uncertainty inherent in these questions. But if I *had* to answer, then, on the basis of its supreme combination of simplicity and explanatory depth, I would proudly declare 'teleological cosmopsychism!' and hope for the best.

What would you say?

Choice Point: If you've heard enough to be able to make sense of the reality of cosmic purpose, feel free to jump to Chapter 7 where we'll explore the implications of cosmic purpose for human existence. Alternately, stick around this chapter to delve deeper into some of the details.

Digging Deeper

Will Physical Science Ever Explain Consciousness?

Whilst most agree that we haven't *yet* fully accounted for consciousness in the terms of physical science, how can I prove that we *never will*? The problem is that you can't fully articulate the qualities of experience in the purely quantitative language of neuroscience. That's not to say you can't pin down *anything* about experience in quantitative terms. Colour experiences, for example, can be fitted into a rich similarity space, along the dimensions of hue, saturation, and brightness, and this can all be captured in mathematical terms. But the language of neuroscience can't fully capture *the qualities* that fill out that structure: the redness of a red experience, for example.

If neuroscience *could* fully capture the qualities of experience, then we would be able to convey to someone blind from birth what it's like to see red by giving them a Braille neuroscience textbook outlining the corresponding patterns of neural firings. But this is absurd, as the concepts employed in neuroscience—essentially concepts of *movement, arrangement, structure,* and *causal connections*—are just radically different from the concepts we employ when we attend to our conscious experience—the concepts of *feels, sensations,* and *qualities.* The advance of neuroscience has done nothing to help the blind grasp the private qualities that fill out the structure of our colour experience, and there is no reason to think it ever will. Interestingly, there is actually an impressive consensus of 60 per cent of Anglophone philosophers who agree that you could never capture the qualities of conscious experience in the purely quantitative vocabulary of physical science.[21] Of course, this doesn't settle anything, but such a rare consensus among philosophers should at least give us pause for thought.

So far, all I have shown is a kind of *descriptive limitation* inherent to physical science, due to the kind of vocabulary physical science works with. But I believe this descriptive limitation entails an *explanatory limitation*. Consider the deep red you experience as you watch the setting sun. Imagine someone claimed to have come up with a brilliant neuroscientific theory that explains why your experience has that specific quality. That theory would first have to *fully articulate* that quality, before showing how it can be accounted for in terms of patterns of neural firings. But no theory framed in the distinctive vocabulary of neuroscience could fully articulate this quality. If you can't even fully describe redness in the language of physical science, then you certainly can't explain it.

We can formulate this argument in two stages:

- Stage 1: You cannot fully articulate the redness of a red experience in the purely quantitative language of physical science.
- Stage 2: If you wanted to fully explain, say, the redness of a red experience in the purely quantitative language of physical science, you'd first have to fully articulate that quality in the purely quantitative language of physical science.
- Conclusion: Therefore, you cannot fully explain the redness of a red experience in the purely quantitative language of physical science.

This is of course just the bare bones of the argument. For more detail, you can read the first half of my academic book *Consciousness and Fundamental Reality*.

We shouldn't be surprised that we can't explain consciousness in the terms of physical science. As I argued in my book *Galileo's Error*, physical science was designed, by the father of modern science Galileo Galilei, to exclude consciousness. Galileo wanted the new science to be purely mathematical, and in order to ensure this, he excluded the qualities of consciousness from its domain of enquiry. This narrowing of the focus of science was a good move, allowing scientists to focus on those aspects of reality that can be captured in mathematics. But if we now want to bring consciousness into the domain of science, we need to rethink the conception of science bequeathed to us by Galileo.

Teleological Cosmopsychism and Pan-Agentialism

How well does teleological cosmopsychism fit with the pan-agentialism defended in Chapter 3? I characterized pan-agentialism as a particle-based theory, on which conscious inclinations are imparted to particles by the wave function. Indeed, on the standard understanding of Bohmian mechanics, physical reality over and above the wave function is made up of particles rather than fields. If we stripped away the Bohmian mechanics, we could bring together pan-agentialism and cosmopsychism by identifying the universe with a universe-wide field, with the result that the universe will behave in a predictable way because the universe is conscious but lacks experiential understanding, and so inevitably acts through the basic rational response: do what you feel like doing. But, contrary to the *teleological* form of cosmopsychism, this would not be a view on which the universe is maximizing the good; it's just doing what it feels like doing.

In order to combine teleological cosmopsychism with pan-agentialism, we need to identify the cosmic fine-tuner with the wave function itself. On the resulting view, the wave function is a conscious entity that is aware of the complete future consequences of the options available to it, and acts by choosing the best one. During the Planck Epoch, the best option available to the wave function was to put itself in a state whereby the universe would become life-permitting. The apparently mechanical behaviour of the wave function thereafter reflects the limited options available to it.

At the end of Chapter 3, I argued that fine-tuning and rational matter fit together like a lock into a key, and that this is unlikely to be a coincidence. On teleological cosmopsychism we start with rational matter, as particles and the wave function are themselves rationally responsive material entities. The wave function then fine-tuned itself in order to allow matter to reach a greater realization of its rational nature. In this sense, the fit between fine-tuning and matter is not a coincidence.

7
Living with Purpose

I have argued that the universe has a purpose. But what is that purpose? We know what it is at least in part. The fine-tuning of the cosmos suggests that the emergence of life is among the goals of the universe. And in Chapter 3, I argued that the emergence of rational organisms—a category in which I include non-human animals—is also part of the cosmic purpose. But what else?

It's possible that that's it. 'That's all folks!' as Porky Pig used to say at the end of cartoons. Maybe this is all the conscious universe or the designer of limited power can manage, or all that is written into the teleological laws. But it seems improbable that the time we find ourselves in happens to be the endpoint and final fulfilment of reality. It is more likely that the purpose of the universe is still unfolding, and that new forms of existence will emerge, as unfathomable to us as our existence is to worms.

How will these new forms of existence emerge? If an alien from another universe had visited a few billion years ago before life emerged and examined the inanimate matter that exclusively filled the universe, observing the mechanical rules governing its behaviour, this alien would never have dreamt that this same stuff would one day achieve self-consciousness, rational understanding, and moral awareness. And yet, that potential was always there in matter, waiting for the right conditions for it to emerge. It is possible that built into the stardust that makes us up is the potential for some even higher form of existence, as yet invisible and unrealized.

What conditions will bring about the next stage of cosmic evolution? In truth, we don't know. But it can be rational to hope beyond what the evidence supports. We have the ability to make ourselves and our world better, and we can choose to live in the hope that by achieving the highest state that is possible for our form of existence (we've got a long way to go!) we will somehow lay the foundations for the next leap forward. We can call this way of living 'cosmic purposivism.'

This hopeful commitment to our capacity to advance the purposes of the universe transforms our ethical situation. True ethics is not about helping your kin alone—the exclusive concern of the Mafia boss—or helping your nation alone—the exclusive concern of the fascist. True ethics is a concern to *make reality better*. If there is no cosmic purpose, then making reality better is mostly a negative project, in the sense that it largely consists in removing bad stuff, such as suffering and injustice. Removing suffering and injustice is incredibly important, and one can live a highly meaningful life as a humanist dedicated to this end. But if cosmic purpose is still unfolding, and if our actions can contribute—even in some small way—to bringing about the next stage of cosmic evolution, then the potential consequences of our actions are so much greater than they would be in the absence of cosmic purpose. We may be able to contribute to bringing about a vastly superior state of existence to the one we currently inhabit.

Even if there is a great state of cosmic evolution yet to come, it's possible our actions are irrelevant to bringing it about. Perhaps the trigger will be some entirely unexpected event. However, the fulfilment of cosmic purpose has thus far consisted in a process of making the universe progressively better: the emergence of life and later intelligent agents. Assuming the next stage continues this story of cosmic progress, our best guess as to how to hasten its coming is by making the world as good as we possibly can. At the very least, we will bring reality closer to that higher state, if only by a little bit.

The ethical project of the cosmic purposivist may therefore be of vastly greater significance than that of the humanist. For sure, the ethical project of the cosmic purposivist is more likely to fail, but this is just a reflection of the fact that the ethical project of the cosmic purposivist is more ambitious than that of the humanist. Similarly, a bold environmentalist who wants, against challenging odds, to reach net zero is less likely to succeed than a timid environmentalist who merely wants to burn a little bit less carbon. But there is something noble about living in hope of a great goal, and fighting for it, even if the odds are somewhat against you.

I don't want to definitively say that the life of a cosmic purposivist is more meaningful than that of a humanist. I can say that I have found living as a cosmic purposivist to be a deeply meaningful form of life. For me, 'prayer' consists in a daily effort to commit to living not only for my own interests, and those of my loved ones, but ultimately for the sake of advancing the good purposes of ultimate reality. This does not mean caring less for family and friends or relinquishing one's personal life goals. Rather it means seeing these narrow interests as part of a broader project that transcends them. I am fortunate not to be very materialistic by nature, but I do have a certain amount of ego and personal ambition. Ego is a powerful force: an insatiable appetite to be the centre of the universe that only grows stronger the more it is fed. I have found that continuously redoubling efforts to live for a purpose greater than myself helps to keep these negative traits in check. Over time, a persistent focus on cosmic purpose has made me a slightly better person.

What more can we say about the project of the cosmic purposivist? Of course, different people will have different views of what 'making reality better' consists in. In what follows, I will offer my own views, starting with values of a spiritual nature before moving to more worldly values.

Spiritual Advancement

When you look around you, you feel as though you're looking *directly* at the world, and seeing things as they are in and of themselves. But a little reflection renders it obvious that we experience reality in a highly culturally conditioned way. This is perhaps most obvious when we reflect on the experience of language. When you hear someone say to you 'Get out! The building's on fire!', you can't help hearing those words as meaning what they do. Suppose you're standing next to a Chinese speaker who doesn't speak English. They would just hear meaningless sounds. In a sense, that would be closer to the truth, to experiencing what's really out there in the world. After all, the sounds themselves are just vibrations in the air. They don't contain any 'intrinsic meaning.' And yet, when you know a language well, you experience sounds as *intrinsically meaningful*. This is just one trivial way in which we project our cultural conditioning onto the world.

This particular form of culturally conditioned experience can be interrupted, at least a little bit, simply by persistently attending to a specific word. If you say the word 'bread' to yourself numerous times, it starts to lose its meaning; you find yourself experiencing meaningless phonemes rather than a meaningful word. This simple game is a useful way of appreciating one of the major purposes of meditation: using persistent attention to break cultural conditioning.

The week before I got married, I spent some time in a monastery to spiritually prepare, spending a few hours every morning meditating. When you first start meditating, it feels like you're just attending to your breathing in an entirely passive way: just experiencing what is there to be experienced. However, after you've focused for a long enough period of time, it starts to become apparent that you're not merely passively receiving what's there to be experienced, but in subtle ways you're projecting onto the experience *an idea of breathing.* It's so subtle that it's hard to put into words what this amounts to, but it's kind of like you're projecting a slight caricature of the experience of breathing onto the experience itself.

Reflecting on this personal discovery (that I'm sure countless people have discovered for themselves) fundamentally altered my perspective on reality. It really brought home to me the extent to which we're doing this all the time without noticing: projecting onto the world an idea of it rather than just experiencing what's there to be experienced. It may even be impossible for us to do the latter: just to passively experience sensory qualities—the feel of the breath, the redness of the tomato—without projecting onto them. Having said that, unconditioned experience is an ideal we can, with effort, get closer to. A white racist may experience people with dark skin as sub-human and homogeneous. Getting to know a variety of non-white people may break this conditioning, revealing the emptiness of his earlier conditioning, and thereby helping him to get a bit closer to the ideal of unconditioned experience.

Art can also help with this. Bad art is banal; it simply follows cultural expectations without doing anything new. The punk movement tried to counter bad art by aggressively breaking all cultural expectations. I'm a huge fan of the original punk bands. But the problem is that this approach is not sustainable. Soon after the original movement, 'punk' quickly became just another established style with its own conventional rules, another culturally conditioned way of experiencing reality. True art is a subtle middle way between succumbing to cultural conditioning on the one hand, and aggressively rejecting it on the other. True art works within culturally conditioned forms but—by making new moves from within those forms—undoes them from within, revealing their inherent emptiness. A perfect example is my fellow Liverpudlians, The Beatles. The Beatles started off with the culturally understood norms of rock and roll and pushed those norms into places they'd never gone before, creating the greatest popular music ever made, before or since.

If art and meditation gently chisel away at our conditioned way of experiencing reality, psychedelics hack off huge chunks in one go. This can be terrifying. But done carefully, taking psychedelics can be incredibly liberating and enlightening. I took psychedelics a lot when I was a teenager. Many psychedelic experiences are hard to communicate, but here's a small example of how they can break cultural conditioning. I remember watching the BBC news and being struck vividly by the absurdity of the dramatic music in the credit sequence, as though it was drumming up excitement in preparation for hearing factual information read out. Previously I had been used to this, it had become for me a 'natural' and inevitable part of how things are. The psychedelics simply removed the culturally conditioned ways of hearing that made me experience this as natural and inevitable, and brought me a little bit closer to seeing things how they really are. And it was very, very funny.

Psychedelics have been used by human beings for spiritual advancement for centuries if not millennia.[1] They are not addictive and their use does not appear to cause physical harm. It's hard to see why they should be lumped together with highly addictive and harmful substances like heroin and cocaine, except for the cynical reason that citizens may be harder to manipulate if they can see through cultural conditioning (I'm convinced Brexit wouldn't have happened if psychedelics were more widely available). There are of course psychological dangers posed by psychedelics; suddenly having your conditioned understanding of reality torn away can be incredibly frightening, and can lead to intense panic.[2] But there are also concrete benefits, with recent studies showing that psychedelics have extraordinary potential to cure people of a wide range of debilitating mental health problems.[3] Psychedelics should be used carefully, in the right circumstances, and after psychological assessment. The best way for society to ensure this is by decriminalizing their use in controlled circumstances. We need to reclaim our right to this powerful tool for spiritual advancement.

The characterization of spiritual advancement I have given thus far might seem unduly negative. What's the point of just breaking cultural conditioning? After all, we wouldn't be able to live normal lives if we literally broke through all of our cultural ways of experiencing the world and just experienced sensory qualities without any cultural meaning. From a spiritual perspective, the big positive attraction is that many who have made significant progress in breaking through their conditioning, perhaps by meditating on their breath until they experience pure bodily sensation, testify that there is a higher form of consciousness underlying our culturally conditioned forms of experience. We call such states of awareness 'mystical experiences.'

Things get even harder to articulate here. The content of a mystical experience is reported to be *ineffable*, meaning its character cannot be expressed in ordinary language. Ineffability itself is not unique to mystical experiences: the character of a red experience is also ineffable, in the sense that it cannot be communicated to someone who has never seen red. However, whereas the ineffable aspect of a red experience just concerns the experience itself, a mystical experience has what the great 19th-century psychologist and philosopher William James called a 'noetic feel,' meaning that it seems to the person undergoing it to be a way of knowing about reality outside of the experience.[4] In a mystical experience one seems to directly encounter a life or living presence that exists in all things. Some call it 'God' or 'Brahman.' In order to stay neutral on many metaphysical questions, James simply called it the 'More.'

If you're a materialist, this experience must be a delusion. According to materialism, the fundamental story of reality is the purely quantitative story we get from physics, and this is not a story that features a 'living presence' at the fundamental level of reality. However, if one is a panpsychist, if one already thinks that all of reality is infused with consciousness, it's not too much of a leap to suppose that the living presence one encounters in mystical experience is an aspect of the consciousness that permeates all matter.

Even if all of this is *consistent* with panpsychism, surely we should require evidence before accepting the bold claim that mystical experiences put us in direct contact with the ultimate nature of reality? To the contrary, James argued that the demand for evidence to 'prove' that mystical experiences correspond to reality when we don't, and indeed can't, require proof that our ordinary, sensory experiences correspond to reality, introduces a pernicious double standard. I agree.

Imagine you wake up at the bottom of a deep, dark hole with total amnesia. You have no idea who you are or how you got here. A voice from the top of the hole is speaking to you, explaining how you got there and what you need to do to get out. Do you have any reason to think the voice is telling you the truth? Without access to your memory, you have nothing to go on in assessing the credibility of the speaker. They could be telling the truth but they could equally be lying. Nonetheless, you have strong pragmatic reason to trust the voice. After all, what else are you going to do?

The above is a good metaphor for life. Each person finds themselves stuck in the 'hole' of their own conscious mind, with no means of escape from its boundaries. But from within this prison, we find ourselves subject to sensory experiences which 'tell' us about a world outside of our minds. My visual experience right now tells me there is a table in front of me with a laptop on it. I cannot climb out of my conscious mind to check whether my experiences correspond to the real world. They could be caused by a physical world around me but they could equally be delusions created by the evil computers running the Matrix. I nonetheless have solid pragmatic grounds for trusting my experiences. After all, what else am I going to do?

Could mystical experiences be delusions? Of course they could. But then so could our sensory experiences. In a particular case we can test the reliability of our senses, but we can do so only by using our senses, thus rendering circular any attempt to prove that sensory experience as a whole puts us in touch with objective reality. All knowledge of the reality outside of our minds is rooted in leaps of faith, in decisions to trust what experience tells us about reality, and there's no good reason to think that faith in one's mystical experiences is any less rational than faith in one's sensory experiences.

Mystical experiences are rare (although would be a lot less rare with the decriminalization of psychedelics). But mystical experiences are simply more powerful forms of what the German philosopher and theologian Rudolph Otto called 'numinous' experiences, in which one has a brief sense of the awe-inspiring, sacred depth of reality.[5] Through meditation and simple living, through engagement with nature and with things of beauty, through persistent effort to break one's conditioned way of experiencing reality, each of us can become a little more deeply acquainted with the 'More,' and in doing so make reality a little bit better.

Spiritual Communities

I was raised Catholic, in a vibrant and flourishing community in Liverpool. At 14, I decided I was an atheist and so refused to get confirmed, upsetting my matriarchal grandmother. My mum sent me to have a chat with Fr Paul Fegan, our charming and youthful parish priest, who tried 'Pascal's wager' on me:

> If you carry on as a Christian and it's all false, you haven't lost much, have you, except the chance to lie in on a Sunday morning. But if you throw it all in, and on the other side you find out God does exist, you're going to feel like a bit of a Charlie, aren't you?' (I distinctly remember he used the word 'Charlie' to mean fool, which I hadn't heard before and haven't heard since.)

I replied that I didn't think a good God would respect me for believing on the basis of this kind of coldblooded calculation. (If I'd read William James at that age, I might have said that faith that is formed in this way 'would lack the inner soul of faith's reality.'[6]) Furthermore, how could I know Christianity was the right religion to bet on? Maybe in the afterlife I'd encounter a very cross Allah annoyed that I believed in the Trinity. At the end of our chat, Fr Paul smiled at me fondly and said I was making the right decision not to get confirmed. My mum wasn't best pleased but respected my decision.

I didn't have much to do with institutional religion for the next twenty years. However, in my early 30s, as my taste for late night drinking began to lessen—these days I only stay up late drinking at philosophy conferences—I began to feel the absence of the things I took for granted in my Catholic upbringing. In secular life, almost all social encounters happen because of what you can get out of the other person. I don't mean that in a cynical way; there's nothing wrong with getting together with someone because you enjoy their conversation, or because you share a passion for playing tiddly-winks. But in the church of my youth, you connected with people just because they were your neighbours. You felt part of a family with people of all ages and all socioeconomic backgrounds. You marked the seasons together. You rejoiced together when someone was baptized or married and wept together when someone died.

Such reflections eventually took me to a service at Liverpool Cathedral shortly before Christmas. Anglican evensong is a beautiful tradition, and the music and majestic beauty of the space took me to a place I hadn't been for some time. I stayed for coffee afterwards and ended up chatting to the Dean of the Cathedral, Pete Wilcox, now Bishop of Sheffield. I told him how much I'd enjoyed the service, but that I wasn't sure how literally I could take the beliefs of Christianity. To my surprise, Pete replied, 'I'm a conservative Christian, I believe Jesus rose from the dead leaving an empty tomb. But there are certainly plenty of Christians who don't take things so literally— including members of this congregation!' In contrast to the rigid belief demands of the Catholicism of my upbringing, this response blew me away.

Not only was I not turned away, but Pete the Dean gave me books to read by liberal religious thinkers. One of these was Marcus Borg, a New Testament scholar and liberal theologian who formulated a conception of Christianity involving few of the beliefs standardly associated with Christianity, such as a literal resurrection and a personal God. Borg affirmed the existence of God, but a God whose nature could not be expressed in human language, and hence who is not literally 'all-knowing' or 'all-powerful.'[7]

This may strike readers as contrary to the 'Christian' idea of God. However, from the very early days of Christianity, there has been a tradition of 'apophatic' or 'negative' theology, according to which God's nature is beyond language. Pseudo-Dionysius the Areopagite (late 5th/early 6th cen-turies) talked of how God is 'beyond every assertion' and 'beyond every denial.' And the late-14th-century text *The Cloud of Unknowing* was hugely influential in showing Christians how to move beyond the descriptions of God found in ordinary worship to a deeper experience of a God beyond human characterization. Even some of the Early Church Fathers, such as Origen (c.184–253) and Gregory of Nyssa (c.335–395), adopted the apophatic approach.

I have been a consistent atheist about the Omni-God since I was 14. However, I have always taken seriously the reality of the 'More' that is known in mystical experiences. I've never had a full-blown mystical experience, but occasionally, after prolonged meditation, deep experiences of beauty, or sometimes during the soft light of dawn or dusk, I have a fleeting sense of a greater reality at the root of things. Our scientistic culture ridicules believing without empirical evidence, but it seems to me perfectly rational to take numinous experiences themselves to be a valid source of empirical evidence. Reading about apophatic Christianity, I came to the startling realization that this conception of 'God' resembled—was perhaps even identical to—something I believed in.

What about the story of Jesus, including its many miraculous occurrences? While there is much history we can get out of the gospels, Borg argued that, from a religious perspective, we should think of the Christian story not as conveying historical fact, but as expressing what he called the 'character and passion' of God. Through meditating on this story, in which God is identified not with the king in his castle but with the naked, executed peasant—the guy who was born in a barn and hung out with the outcasts of society—we are afforded a deep insight into what God—or the 'More'— truly is. For Borg, the resurrection was not about a corpse coming back to life, but about the transcendent reality he knew through the character of Jesus still being alive and active in the world.

In other words, the Christian story is understood not as literal fact but as a profound fiction, one that, as part of the Christian spiritual practice, facilitates a deeper connection with ultimate reality. I had found what I was looking for: a way of connecting to my community, my tradition, and to the purpose of existence, consistent with my lack of belief in traditional Christianity. Borg's view fits with the religion of my culture: Christianity. However, the philosopher John Hick defended a similar conception of religion to Borg but broadened to encompass all religions.[8] For Hick, all religions are connecting with the More—Hick called it the 'Real'—but doing so with culturally specific mythological language. This general approach to treating religion as important fiction is known as 'religious fictionalism.'[9]

My journey back to traditional religion was perhaps a little unusual in that it didn't have much to do with belief, at least not the beliefs of traditional Christianity. This contrasts with the common assumption that religion is all about belief. We often refer to religious people simply as 'believers.' But there is more to a religion than a set of doctrines. Religions involve spiritual practices, traditions that bind a community together across space and time, and rituals that mark the seasons and the big moments of life: birth, coming of age, marriage, death. This is not to deny that there are specific metaphysical views associated with each religion, nor that there is a place for assessing how plausible those views are. But it is myopic to obsess about the 'belief-y' aspects of religion at the expense of all the other aspects of the lived religious life.[10]

For tens of thousands of years, religion has played the social role of binding communities together and connecting them with the More. Now, at least in the US and Western Europe, participation in traditional religion is in decline.[11] There are many reasons for this, but one reason is an intellectual climate in which it is held to be irrational to believe the doctrines associated with traditional religion. And if religion is all about belief, it follows that traditional religion is irrational: a 'delusion,' you might say.

On the other hand, spirituality has by no means died out. Even in the very secular UK, a recent survey found that 46 per cent of UK adults agreed that 'all religions have some element of truth in them,' and 49 per cent that 'humans are at heart spiritual beings.'[12] But in the absence of belief in traditional religion, many of these people become 'spiritual but not religious.'

I'm not here to defend the One True Path for all people. I'm sure being 'spiritual but not religious' works really well for many people. But there are advantages to belonging to a religious community. Spiritual practice is hard. It requires discipline, struggle, wrestling with vice and human frailty. It can help to have the support of a loving community engaged on the same path, and the resources and structure of a rich religious tradition that stretches back thousands of years. My aim here is simply to make space for the 'spiritual but not believers' to take advantage of what traditional religion has to offer, if they choose to do so.

Contemporary religion takes many forms. An option for those seeking the benefits of religion who don't feel comfortable being part of a religion associated with specific doctrines—even if they don't have to believe those doctrines—are the more or less creedless religions, such as the Quakers or the Unitarians. And there will, of course, always be spiritual people who, for whatever reason, prefer to keep their spiritual practice personal. All of this is consistent with my conviction that society would benefit from the majority of people feeling able, culturally and intellectually, to engage with religion, should they want to.

Margaret Thatcher was wrong when she said, 'there is no such thing as society.' We are not just isolated individuals. Through our culturally conditioned way of seeing the world we are bound together in a shared form of life. Spiritual advancement, therefore, is not just a matter of individual spiritual practice. It is rather a matter of building spiritual communities, shaping a society in relationship with the More, a relationship that can deepen over time. This is simply not possible without religion of some kind.

Owning and Belonging

Throughout much of the history of human civilization, a very small number of the population have hoarded all of the wealth and power for themselves. In Europe in 1913, the top 10 per cent owned 89 per cent of all private property, whilst the entire bottom half of society owned just 1 per cent. However, during the 20th century, a remarkable, and historically unprecedented, transformation took place: a middle class emerged. That is to say, a significant proportion of the wealth at the top was transferred to the middle 40 per cent between the top 10 per cent and the bottom 50 per cent. In 2020, this middle group owned about 38 per cent of all property, whilst the share of the elite 10 per cent at the top had significantly come down from 89 per cent to 56 per cent. The bottom half of society benefited a little but not a lot from this transition: moving up from owning 1 per cent to owning 6 per cent.[13]

This transition was not due to capitalism, which had already existed for hundreds of years. It was due to the specific form of capitalism that emerged in the post-war years, which involved democratic control of the market, through active trade unions, strong regulation on the movement of capital, and high taxes on the wealthy. The top marginal rate of income tax in the US between 1932 and 1980 averaged at 81 per cent. In the UK during the same period, the average of the top rate of tax was 89 per cent. These taxes did not dampen economic activity.[14] On the contrary, this time was referred to as the 'global age of capitalism,' as the US and Western Europe enjoyed rates of average growth not experienced since. The fact that high taxes on the wealthy succeeded in making society much more equal without negatively impacting the economy is not often mentioned by political pundits and commentators. It seems to have been wiped from the collective consciousness.

Despite the benefits, both in terms of growth and making society more equal, this model of restrained capitalism was abandoned in the Thatcher/ Reagan 'neo-liberal' revolutions of the early 1980s. Regulation and taxes on the wealthy were slashed, with the promise of unleashing economic dynamism that would produce wealth that would eventually 'trickle down' to the masses. Instead, we got growing inequality, lower growth, and ultimately the global meltdown of 2008. Perhaps the architects of this change can be forgiven for thinking that Wild West capitalism would yield better results. But we've had the experiment and the results are in. On pretty much any measure, democratically constrained capitalism, with higher taxes on the wealthy, works better.

Of course, the social progress enjoyed in the US and Western Europe after the war was not shared by most of the rest of the world. Historically, Western European countries sucked wealth from other nations through military conquest and empire. Before Britain took control of India in the 17th century, India's GDP had consistently been between 25 per cent and 35 per cent of global GDP for over 1,500 years. When Britain left India in 1947, India's GDP was 2 per cent of global GDP. Strictly speaking, European empires have now been largely dismantled. However, rich countries continue to steal from poorer countries through the more nefarious means of tax havens. A multinational corporation can profit in, say, Nigeria, but avoid paying tax in Nigeria by shifting profits to a tax haven. This is commonly done by the mechanism of transfer pricing, whereby the Nigerian branch of the corporation will buy products from the sister company located in the tax haven at ridiculously high prices, thereby moving profits made in Nigeria to, say, the Bahamas. The amount of money lost to the developing world from outflow of multinationals' profits and capital flight is several times greater than the money rich countries spend on international aid.[15] This is just one way in which wealthier countries use their economic power to keep many poor countries in a state of permanent impoverishment.

What can be done? One of the biggest barriers to change is the myth of natural, inviolable property rights. The wealthy resent the kind of taxes that did so much good in the post-war years, because they feel that the taxes would be taking 'their money' off them. In the most extreme version of this sentiment, fans of the 20th-century Russian-born American pop-novelist Ayn Rand declare that 'taxation is theft,' at least if it is used for anything more than funding the police to protect property rights. But even politicians on the centre-left tend to start from the assumption that the need to fund public services or—heaven forbid—redistribute income must be balanced against a concern about the state taking too much of 'our money'. In this climate, the poor will vote to protect their morsel from the nasty tax collector, while the wealthy get to keep their billions.

The conviction that the money I get in my pay-packet before the deduction of taxes is, in some morally significant sense, 'mine', although almost universal, is demonstrably confused. As we discuss in the appendix of this chapter, 'P.S. Is Taxation Theft?', there is no serious political theory according to which my pre-tax income is 'mine' in any morally significant sense. Nonetheless, the illusion of having a sacred, inviolable right to one's pre-tax income is hard to shift. It's a deep part of our socially conditioned way of seeing reality. Anyone who has raised children knows how early and with what force they begin to declare 'Mine!' This is one reason why the political struggle for justice is inseparable from the spiritual struggle to break our social conditioning. In our quest for spiritual advancement, we need to see the myth of sacred property rights as part of the enemy we're battling, both in ourselves and in others.

There are similar psychological difficulties engaging with the rights of non-human animals, and of plants and trees. Sometimes I wonder whether a mass encounter with the living presence in all things, brought about through widespread safe and legal use of psilocybin, might be the only way to combat the ongoing commodification of nature which is launching us headlong into climate catastrophe. Here again, spiritual advancement and political progress could go hand in hand.

What economic model fits with a recognition of the constructed and transient nature of property rights?[16] The French economist Thomas Piketty has formulated detailed plans for what he calls 'participatory socialism', a system in which ownership is understood to be temporary, in which wealth and power continuously flow through the economy, bringing both liberation and true dynamism.[17] To take just one of his proposals, a high wealth tax would fund a 'universal inheritance', meaning each citizen receives 60 per cent of average wealth on their 25th birthday. In the US, this would be $150,000; in Europe around 120,000 Euros. In contrast to the reckless revolutionary zeal of Soviet communism, Piketty talks of a gradual 'decommodification' of the economy. It would be unwise to try to do away with capitalism overnight. But a hopeful, spiritual vision for the future of humanity is inconsistent with the idea that we will never move beyond a profit-driven economy.

What I would hope for is a movement that is both political and spiritual: fighting for a just society not simply because it is good in itself but as a part of a great hope for the future of the universe.

The Purpose of the Universe

As a society, we have got used to the idea that science has ruled out any kind of cosmic purpose. I believe this common conviction has two roots:

- A lag in culture keeping up with changing evidence. Over a hundred years with no evidence for cosmic purpose embedded in our culture the idea that science has ruled out cosmic purpose, and it will take time for the culture to change in the light of the scientific evidence we now have for cosmic purpose.
- Our adoption of a scientific approach which focuses exclusively on public observation and experiment, at the cost of ignoring other sources of data, in particular the privately known reality of consciousness. The universe looks very different when we stop pretending we're mechanisms.

I am confident that this will change in the decades to come. We can only ignore the evidence of cosmological fine-tuning for so long. And my sense is that the public are ceasing to believe that the mind-body problem is just a scientific problem we are one experiment away from solving. A radical change in how we see the universe is on its way.

At the same time, we are living through a scary, uncertain era. Nothing has filled the vacuum left by the decline of traditional religion. And whilst the 2008 global meltdown gave the lie to the delusion that Wild West capitalism will provide peace and prosperity for all, no political philosophy has come along to replace neo-liberalism.

My hope is that cosmic purposivism may point the way to a new optimism in human potential, a faith based not on dogmatic certainties but on a humble and open exploration of an unfolding purpose we don't yet fully understand. Times of big change can be frightening, but they are also pregnant with opportunities for renewal. We have every reason to feel optimistic about the future.

P.S. Is Taxation Theft?

I claimed in Chapter 7 that 'there is no serious political theory according to which my pre-tax income is "mine" in any morally significant sense.' In this appendix, I explain why, starting with the question, 'Is taxation theft?'

Desert or Entitlement?

In addressing the question of whether taxation is theft, it is important to distinguish two senses of 'theft': legal and moral. In 18th-century North America, it was possible to 'own' a slave, in the legal sense of ownership. If someone deprived me of my slave in order to give that slave liberty, then this constituted 'theft' in the legal sense. But of course the laws underpinning slavery were morally abhorrent, and hence few these days would class liberating a slave as 'theft' in any moral sense. Conversely, we can have cases of moral theft that are not legal theft. The laws of Nazi Germany enabled the authorities to seize the property of Jews who had been deported; although strictly speaking legal, such actions constituted 'theft' in a moral sense.

And so, when we ask ourselves whether taxation is theft, we have to specify whether we are thinking of the moral or legal sense (or both). If we wanted to say that tax is legal theft, then we would have to argue that people have a legal claim to their pre-tax income, and hence that the government commits legal theft when it takes the pre-tax income of its citizens. This idea can be quickly dismissed. Clearly if Ms Jones is legally obliged to pay a certain amount of tax on her gross income, then she is not legally entitled to keep all her pre-tax income. It follows logically that the state does not commit legal theft when it enforces the payment of this tax.

The more interesting question is whether taxation is moral theft, and this depends on whether citizens have some kind of moral claim on their gross income. It is to this question I now turn.

Your gross, or pre-tax, income is the money the market delivers to you. In what sense might it be thought that you have a moral claim on this money? One answer might be that you deserve it: you have worked hard and have done a good job, and consequently you deserve all your gross income as recompense for your labour. According to this line of reasoning, when the government taxes, it takes the money that you deserve for the work you do.

This is not a plausible view. For it implies that the market distributes to people exactly what they deserve for the work that they do. But it's not plausible that a hedge-fund manager deserves many times more wealth than a scientist working on a cure for cancer, and few would think that current pay ratios in companies reflect what philosophers call desert claims. Probably you work very hard in your job, and you make an important contribution. But then so do most people, and the market distribution of wealth patently does not reward in proportion to how hard-working people are, or how much of a contribution they make to society. If we were just focusing on desert, then there is a good case for taxation to correct the amoral distribution of the market.

If we have a moral claim on our gross income, it is not because we *deserve* it, but because we are *entitled* to it. What's the difference? What you deserve is what you ought to have as a result of hard work or social contribution; what you are entitled to is the result of your *property rights*. Libertarians believe that each individual has *natural property rights*, which it would be immoral for the government to infringe. According to right-wing libertarians such as Robert Nozick and Murray Rothbard, taxation is morally wrong not because the taxman takes what people deserve, but because he takes what people have a right to.[1]

Therefore, if taxation is theft, it's because it essentially involves the violation of people's natural rights to property. But do we really have natural rights to property? And even if we do, does taxation really infringe them? To begin to address these questions, we need to think more carefully about the nature of property.

Is Property Theft?

The French anarchist Pierre-Joseph Proudhon declared in 1840 that all property is theft. But even among those who accept the legitimacy of property, there are very different views as to what exactly the right to property amounts to. Libertarians believe that property rights are natural, reflecting basic moral facts about the world. Others hold that property rights are merely legal, social constructions, which are created by us and can be shaped to suit our purposes. We can call the latter view 'social constructivism' about property (not to be confused with constructivist views about morality in general).

To bring out the difference, ask yourself: 'Which comes first: facts about *property* or facts about property *law*?' For the social constructivist, the right to property is not some natural, sacred thing that exists independently of human conventions and legal practices. Rather, we create property rights, by setting up legal institutions to ensure that people have certain legal rights over the material world. For the libertarian, in contrast, facts about property exist independently of human laws and conventions, and indeed human laws and conventions ought to be moulded to respect the natural right to property.

This distinction is crucial for our question. Suppose we accept the social-constructivist view that property rights are merely legal. Now we ask the question: 'Do I have a moral claim on the entirety of my pre-tax income?' We cannot argue that I am entitled to my pre-tax income on the basis of my natural property rights, as there are no such things as 'natural' property rights (according to the social-constructivist position we are now considering). So, if I have a moral claim on my entire pre-tax income, this must be because it is exactly the amount of money I deserve for my hard work and social contribution, presumably because in general the market delivers to each person exactly what they deserve. But we have already concluded that this is not a plausible claim. Without the belief in natural property rights, existing independent of human laws and conventions, there is no way to make sense of the idea that the deliverances of the market are inherently just, and hence no way to make sense of the idea that each person's gross income (which is just the income the market delivers to them) is hers by right.

Here's where we're up to: to make sense of the idea that taxation is (moral) theft, we have to make sense of the idea that each person has a moral claim on the entirety of her gross income, and this can be made sense of only if property rights are natural rather than mere human constructions. We need, therefore, to defend a theory of natural property rights. Our next task is to explore philosophical theories of property rights.

Three Theories of Property Rights

We can group philosophical theories of property rights into three categories: right-libertarian, left-libertarian, and social constructivist. Let's take each in turn.

All libertarians hold that an individual has full natural rights of ownership over herself, and the fruits of her labour. However, libertarians of the right and the left disagree over the ownership rights that individuals can have over the natural world, i.e. over land and natural resources.

Right-Libertarians

Right-libertarians believe that the material world—all land and everything that stands on it—was once owned by nobody. The first individuals who discover, or claim, or 'mix their labour' (a phrase introduced in this context by the 17th-century philosopher John Locke[2]) with things in the natural world come to possess an inalienable natural property right over those things. If I'm the first to find some land and I farm it, I come to have natural property rights over that land, so that it is morally wrong for someone to take the land or its produce away from me without my consent.

We can now start to see how someone might try to make sense of the view that taxation is moral theft. If we think of the market as a free and consensual exchange among individuals of things over which they have natural property rights, then any state interference with the market will constitute infringement of individuals' natural rights. Taxation will take from citizens what is theirs by right.

Left-Libertarians

Less familiar in popular discourse is the view known as 'left-libertarianism', defended, among others, by Peter Vallentyne, Hillel Steiner, and Michael Otsuka.[3] Left-libertarians agree with right-libertarians that each individual has full rights of ownership over herself and the fruits of her labour. However, they hold that the natural world belongs to everyone: it is not possible for one individual to acquire exclusive rights over land or natural resources in a way that excludes the equal moral claims of other citizens.

There are various forms of this view. In a more extreme version, the natural world is jointly owned by everyone, such that no one is permitted to own anything without the express consent of every other living individual (compare: if we jointly own a house, you can't just let out half of it without my consent). In a more modest form, no one is allowed to acquire property unless they leave enough for each person to have an equally valuable share of natural resources. The uniting principle of left-libertarians is that each of us has an equal moral claim over the resources of the world.

Left-libertarianism will certainly rule out some forms of taxation as immoral. If I have acquired land or natural resources in a way that is consistent with the equal moral claim of others, and through my own labour I increase the value of those resources, it is wrong for the state to tax that wealth away from me. But left-libertarian theories leave considerable latitude for the state to alter the distribution of wealth, perhaps through taxation, if some take more than their fair share of land or natural resources. Crucially, the claims of future generations must also be taken into account, leading naturally to an inheritance tax (or at least restrictions on the right to bequeath) to ensure that each future individual has a fair share of natural resources.

Social Constructivists

As already discussed, social constructivists do not deny the existence of property rights, rather they take them to be social or legal constructions, which humans are free to shape to reflect what they deem valuable. Jesus declared that 'The Sabbath was made for man, and not man for the Sabbath.' Analogously, for the social constructivist, property rights are made to serve human interests and not vice versa.

It is plausible that human flourishing requires certain legally protected rights to property, and hence most social constructivists will advocate a system of property rights. At the same time, there are other things of value—perhaps equality, perhaps reward for hard work and/or social contribution (which as we have seen are not well protected by the market)—and in order to promote these other values, most social constructivists propose making property rights conditional on the payment of taxes. In the absence of pre-existing natural property rights, there is no moral reason to respect the market distribution of wealth (there may of course be *pragmatic*, economic reasons, but 20th-century history teaches that high taxes on the wealthy are quite compatible with a healthy economy).

What Do I Have a Right to?

Almost all politicians and voters start from the assumption that each citizen has some kind of moral claim on her gross income. In fact, we have seen that making sense of this requires some hefty and highly contentious philosophical theses. It requires accepting the general libertarian commitment to property being natural and not dependent on human laws or conventions. And it also requires denying the left-libertarian view that each of us has an equal moral claim on the resources of the natural world.

The second requirement—the denial of equal rights over the natural world—is particularly implausible. On the right-libertarian view, it is perfectly morally acceptable for one person to claim a vastly unequal proportion of land and resources for himself, resulting in his propertyless neighbours being forced to work for him to avoid starvation. By what right can the natural world be appropriated in this way? It is one thing to say that one has exclusive natural rights over *oneself*, but how can we justify exclusive rights over the natural world? And if it can't justify this, right-libertarianism falls at the first hurdle.

The absurdity is even plainer when we reflect on the actual reality of land ownership. Half of England is owned by just 1 per cent of the population. According to research by author and campaigner Guy Shrubsole, individual homeowners hold just 5 per cent of the country's land, while the aristocracy owns 30 per cent. If land in England were divided up equally, each of us would possess almost an acre—roughly the size of Parliament Square—in reality, however, many people do not even have a house to live in.[4] Most of the English aristocracy possess the land they do simply because of who their ancestors were friends with. It is very hard to see how this could be the basis for any kind of moral entitlement.

Moreover, *even if* right-libertarianism is true, *even if* there are natural property rights, *even if* such rights allow private individuals to carve off for themselves a vastly unequal share of natural resources, even then we cannot make sense of the idea that actual people living today have a moral claim on their pre-tax income.

The reason is that the world that right-libertarianism theorises about is a very different one to the world we live in today. (It is no accident that the classic academic defence of right-libertarianism—Robert Nozick's 1974 book—is called *Anarchy, State and Utopia*.) According to right-libertarianism, the market distribution of wealth is morally significant because it is the distribution that respects the voluntary choices people have made with the property to which they have a natural right. But this is the case only if the market is *perfectly free*, i.e. if the state has no influence on the distribution of wealth. Yet there are very few countries in the world in which this is the case. In almost every country, there is a certain amount of taxation, at least to pay for roads and infrastructure, if not for education and healthcare. But even the smallest such state intervention entails that the market distribution of wealth no longer reflects the free choices of citizens, and hence *by the lights of right-libertarianism* the citizens of these countries have no moral claim on their pre-tax income.

The point can be made clearer with some examples. Consider a hypothetical Professor Schmidt, a right-libertarian academic working in a German university, who is very annoyed about the state taking 42 per cent of 'her' income. Where did her salary come from? Well, German universities are publicly funded, and so Schmidt's salary comes from general taxation, from the money the German state forcibly extracted from its citizens. But according to right-libertarianism, this is an immoral state action that infringes the natural rights of its citizens; in effect, it steals from people to pay Professor Schmidt. It follows that Professor Schmidt has no right to her salary, and hence no right to complain that the state lets her have only 58 per cent of this stolen money.

Perhaps some radical libertarians will gleefully agree with me that professors who leech off the state have no right to resent taxation. But the point applies quite generally, although in a more subtle way. Now consider a hypothetical Ms Jones, a libertarian British businesswoman who resents paying tax on dividends from her lucrative company. Although she is not directly paid by the state, the profits generated by Jones's business are dependent on many things that are funded by the state: perhaps she receives state subsidies, but even if not, certainly the success of her company will depend on infrastructure, roads, rule of law, and an educated and healthy workforce. It doesn't matter whether in principle these things could have been provided privately; in reality, they are provided by the state and funded through taxation. According to right-libertarianism, these things were paid for by theft, and hence Jones has no right to the profits thus generated.

In theory, right-libertarianism does entail that people have a moral claim on their pre-tax income, and hence that taxation is theft, but only in *purely hypothetical societies where there is zero or minimal state interference in the economy*. In states in which the government intervenes in the economy through taxation—i.e. in almost every developed state—market transactions are tainted and so are morally void. The right-libertarian is perfectly entitled to campaign for the day when her minimal-government Utopia is brought about, but until that day she cannot consistently argue that she has a right to her pre-tax income, and hence cannot consistently complain that the government is taking what is hers by right.

It's hard to shake the feeling that the gross income figure on your payslip represents *your money*, and that the difference from your take-home pay represents how much the state has taken from you. In fact, there is no coherent way of justifying this conviction. Even if the most radical forms of right-libertarianism are true, it remains the case that you have no special moral claim on your gross income in the real world.

Still, the vast majority happily vote for low taxes, rejoicing that they get to keep their morsel while in reality all they've done is protect the spoils of a tiny minority at the top. The result is our failure to create what we really need: a tax system that—as part of the wider economy—creates a just society.

Endnotes

Chapter 1: What's the Point of Living?

1. I am here merely describing Craig's view. In fact, there are scientific theories that imply an infinite number of consecutive universes in the future, for example, Roger Penrose's Conformal Cyclical Cosmology. It is important that we don't conflate *the universe having a purpose* with *the universe being eternal*. In Penrose's cosmology, physical reality of some form or other extends infinitely into the future, but without there being any kind of cosmic purpose. Conversely, we can imagine a universe with cosmic purpose directed towards a climactic fulfilment, after which it passes gracefully out of existence. This book is focused on cosmic purpose, not whether the universe will last forever.

2. See Craig's video 'The Absurdity of Life without God.' https://www.youtube.com/watch?v=ZqNTT0E_T70

3. This argument is made in Benatar (2017a). Benatar is probably better known for the argument for anti-natalism expressed in his 2006 book *Better Never to Have Been*, but I'm not focusing on this argument here, as it doesn't have much to do with cosmic purpose.

4. News story from 7th February 2019, 'Indian Man to Sue His Parents for Giving Birth to Him.' https://www.bbc.co.uk/news/world-asia-india-47154287

5. Benatar (2017b).

6. Teilhard de Chardin (1955).

7. Alexander (1920).

8. Note to academic philosophers: I am here using 'meaning subjectivism' to refer to desire-satisfaction accounts of meaning in life. There are, of course, many other forms of 'subjectivism' which may resist the counting grass concern raised below.

9. One fact about Hume that is not discussed enough, however, is that he was also a horrific racist. Convinced of the superiority of whites over other races, he remarked in a footnote, 'In JAMAICA, indeed, they talk of one negroe as a man of parts and learning; but 'tis likely he is admired for very slender accomplishments, like a parrot, who speaks a few words plainly' (Hume 1753/1882/1964). It is an important question whether we should continue to discuss the work of people from history with such abhorrent attitudes, when racism, both structural and explicit, remains a cause of such deep suffering to many. I'm open to correction on this issue, but my honest view is that it will advance the cause of anti-racism more if we continue to discuss David Hume's philosophy whilst frequently reminding ourselves of and reflecting upon his abhorrent views on race, than if we simply stopped discussing his work at all. It would be a different matter if

Hume's philosophy were infected by his racism, but I don't personally believe that to be the case.

10. Some (Smith 1994) interpret Hume as a non-cognitivist, but I am persuaded otherwise by Pigden (2009) and Olson (2014).

11. Perhaps the subjectivist could say that Susan's life is meaningful relative to her own valuations but meaningless relative to the valuations of others who don't value counting blades of grass (I am grateful to Bart Streumer for raising this point). This is a more radical form of subjectivism, as there is no longer a fact of the matter as to whether Susan has a meaningful life. Moreover, the counterintuitive implication being pressed in this thought experiment merely needs to be put a little differently. Rather than implying that Susan's grass counting is just as meaningful as advancing scientific knowledge, this form of subjectivism implies that Susan's conception of what constitutes meaning is just as reasonable as that of someone who values advancing scientific knowledge.

12. Most theories that attempt to ground value facts in desires (Williams 1979, 1989; Schroeder 2007; Goldman 2009) will do so in terms of an idealized version of the individual's desires, i.e. the desires that would be had if the individual was fully informed and perfectly rational. We can stipulate that an idealized version of Susan's desires would contain a strong desire to count blades of grass. Alternately, one might try to ground the meaning of the individual life in the desires of the group; in response we could consider societies of people who all desire to count blades of grass.

13. Does Susan *desire* counting blades of grass? One problem is that the word 'desire' is used in at least three different ways in philosophy: (A) a conscious yearning for something, (B) having something as a psychological goal, whether or not you yearn for it, (C) enjoying something when you get it, whether or not you consciously yearn for it or have getting it as a psychological goal. Susan 'desires' counting blades of grass in the first sense, which is the relevant sense for the form of subjectivism currently under consideration.

14. Autistic people commonly engage in 'stimming': kinds of repetitive behaviour, such as arm or hand flapping, which non-autistic people may perceive as purposeless. However, these behaviours are engaged in for instrumental reasons. For example, they typically provide pleasure, stress relief, and help to stimulate or reduce sensory input: https://www.autism.org.uk/advice-and-guidance/topics/behaviour/stimming/all-audiences. Hence, these activities are not analogous to Susan's blade of grass counting, which, by stipulation, is done for its own sake. I'm grateful to Roslyn Malcolm for advising me on how to think about this issue in relation to autism.

15. I am imagining an objector who denies that *anybody*—including members of an intelligent alien species—could have the life goal of counting blades of grass. A more modest objection would be that no *human being* could have such a goal, due to the contingent facts about our evolution. In response to this

objection, we could imagine that Susan is an alien. I'm grateful to Mark O'Brien for pointing out this important distinction.

16. I am jumping back and forth between meaning subjectivism and value subjectivism. In principle, one could be a meaning subjectivist without adopting the more general thesis of value subjectivism, but, for the reasons I am about to outline, I ultimately came to reject value subjectivism *per se*.

17. Note to academic philosophers: The counting grass worry is more of a first-order objection against meaning subjectivism (or more precisely desire-satisfaction models of meaning in life), whilst the objection we are about to discuss concerns a certain kind of meta-ethical view. I'm grateful to Chris Cowie for helping me to appreciate this.

18. Hume (1739/2000: 3.1.1, p. 302).

19. Hume (1739/2000: 2.3.3, p. 266).

20. In the broad sense of 'naturalism' defined here we can include expressivists, who think that value claims are expressions of our attitudes. The *Stanford Encylopedia* entries on 'Moral Naturalism' and 'Cognitivism vs. Non-Cognitivism' give a great overview of all of this literature.

21. In fact, many naturalists will accept the Is-Ought Gap principle as defined above, which concerns what we are entitled to *rationally infer* rather than what holds *in reality*, whilst taking that to be compatible with value facts emerging from non-value facts. In other words, we can distinguish an *epistemological* Is-Ought Gap principle (you can't move in reasoning from non-value facts to value facts) from an *ontological* Is-Ought Gap principle (value facts cannot emerge from non-value facts). I have argued extensively in my book *Consciousness and Fundamental Reality* against the kind of separation between epistemology and ontology that would be implied by accepting the former but not the latter (in the context of discussions of consciousness rather than value). I hope in future work to more explicitly connect the framework of that book to the issues about the nature of value discussed in this chapter.

22. Note that we have seamlessly moved through discussions of different sorts of value: meaning, morality, and then value in general. Whilst there are important differences between all of these categories, it was coming to adopt a very general form of ontological (see last endnote) Is-Ought Gap principle which led me to the dichotomy between all-encompassing value fundamentalism and all-encompassing value nihilism.

23. One might object that in the context of this kind of practical deliberation, talk of 'reasons' just expresses claims about the most effective strategy for achieving one's desires. This would be a form of (analytic) naturalism about prudential value rather than the value nihilist position I'm discussing here. But, for what it's worth, it seems to me that this would amount to *redefining* what we usually mean by these terms. We could of course choose to use 'reasons' talk in this way. But, equally, we could choose to use 'reasons' talk in such a way that 'Susan has reason to do *X*' just means 'Susan's doing *X* will maximize human

well-being.' In either case, I don't believe it captures what ordinarily goes on in ethical or prudential deliberation. As a first pass at arguing for this, we could press Moore's famous 'open question' argument (1903: 10–17/1993: 62–9). On the above analysis of reasons talk, it wouldn't make sense to say, 'Doing X would be the most effective way of achieving my desires, but do I really have reason to do it? Maybe all of my desires are misguided, or focused on totally pointless goals'; but this sentence does make sense, and therefore analytic naturalism about prudential value cannot be correct. Obviously, there's lots more to be said.

24. Camus (1942).
25. Streumer (2017).
26. Benacerraf (1973).

Chapter 2: Why Science Points to Purpose

1. There is in fact some controversy about this. For more discussion, see section D of the following blog post by cosmologist Luke Barnes: https://letterstonature.wordpress.com/2015/11/04/neil-degrasse-tyson-on-newton-part-1/

2. Paley (1802). In fact, although Paley used this analogy to make the case vivid, his argument was Bayesian (the form of argument we'll encounter in this chapter) rather than an argument from analogy.

3. Darwin (1859).

4. Dawkins (1986).

5. Nietzsche (1881/2001); Marx (1843/2009); Freud (1927/8).

6. All of the following are good introductions to this data: Leslie (1989); Rees (2000); Davies (2006); Lewis & Barnes (2016).

7. This number actually refers to the proportion of energy that is released when hydrogen fuses into helium, but it roughly corresponds to the strength of the strong nuclear force. This is one of Martin Rees's (2000) '6 numbers.'

8. Lewis & Barnes (2016: 50).

9. Pubchem.ncbi.nlm.nih.gov.

10. Lewis & Barnes (2016: 51).

11. More precisely, we can say the constant is fine-tuned for life if and only if the range of possible values of that constant which permit life are much narrower than the range of possible values that don't.

12. This value of the cosmological constant is taken from Brian Greene's 2012 TED talk 'Is Our Universe the Only Universe?,' https://www.ted.com/talks/brian_greene_is_our_universe_the_only_universe?language=en

13. For accessible introductions to this problem, see Than (2018) and Wolchover (2018).

14. This example, in a slightly modified form, was borrowed from a video by Justin Brierly: https://www.youtube.com/watch?v=yy6kaDaeDT8

15. Barnes (2020).
16. For scientists defending a multiverse account of fine-tuning, see Susskind (2005), Greene (2011), and Tegmark (2014); for philosophers, see Leslie (1989), Smart (1989), Parfit (1998), and Bradley (2009).
17. This original formulation of the inverse gambler's fallacy charge was Hacking (1987); White (2000) is a more developed formulation. I found Draper et al. (2007) especially persuasive. See also Landsman (2016).
18. Goff (2021a).
19. Draper (2020).
20. For a long time, I thought there were obvious counterexamples to the Requirement of Total Evidence, and that the principle would have to be modified to accommodate these counterexamples. I thought, for example, that we couldn't get to the Darwinian hypothesis from the specific evidence that *these particular animals*—the particular animals we happen to observe—came to exist. I wrestled for a couple of years with how to modify the principle to accommodate this kind of counterexample. However, Draper et al. (2007) show how we can use the probability calculus to separate out stronger and weaker evidence, and it turns out that the fact that these particular animals came to exist does raise the probability of the Darwinian hypothesis, whilst the fact that our universe is fine-tuned does not raise the probability of the multiverse hypothesis. This rocked my worlds, and since reading this paper, there is very little doubt in my mind that the inference from fine-tuning to a multiverse is fallacious.
21. Guth (1981, 2000).
22. Adams (2002).
23. Leslie (1989: 13f). I have presented the analogy here in a slightly modified form.
24. Having said that, there are good philosophers who press this objection, e.g. Sober (2003, 2009).
25. Smolin (1997).
26. There are challenges here with how one carves up possibilities into ones to which we should ascribe equal probability. This is one instance of the more general 'reference-class problem' in probability theory, the problem of deciding what class to use when calculating the probability applicable to a particular case. I believe that the thesis that some ways of carving up reality are more *natural* than others has to be part of the solution here. In his book *Writing the Book of the World*, Theodore Sider (2011) has made a more general case for the role of naturalness in science and philosophy. Along these lines, in his articulation of a fine-tuning argument, Robin Collins (2009) formulates the following restricted version of the Principle of Indifference: 'when we have no reason to prefer one value of a variable p over another in some range R, we should assign equal epistemic probabilities to equal ranges of p that are in R, given that p constitutes a "natural variable," where a natural variable is one that appears in the simplest rendering of physics.' The Principle of Indifference may cease to be

applicable in cases where there are multiple, equally natural ways of carving up the possibilities, which would lead to inconsistent results if we applied the Principle of Indifference under each of these carvings. But just because the Principle of Indifference does not apply in *all* cases, it doesn't follow that it doesn't apply in *any* cases. For more on this, see the discussion of 'Classical Probability' in the entry on 'Interpretations of Probability' in the *Stanford Encyclopedia of Philosophy* (Hájek 2019), which summarizes its discussion of the Principle of Indifference by saying 'problems caution us that there are limits to the principle of indifference... [b]ut arguably we must just be careful not to overstate its applicability.'

27. McGrew et al. (2001); Colyvan et al. (2005).
28. Collins (2009).

Chapter 3: Why Consciousness Points to Purpose

1. I am thinking here of the philosophical position which has become known as 'illusionism' (Frankish 2016, Kammerer 2022). Illusionists typically say they believe in 'consciousness' in some sense, but when they say this they mean something entirely defined in behavioural terms (including the behaviour of the organism's internal parts). This contrasts with the idea of 'consciousness' as something that is privately known, and not fully accessible to public observation experiment, which illusionists do not believe in.

2. Why does this not contradict my earlier claims that consciousness is not publicly observable? The sadness of my child is not literally located on her face, and so is not literally an observable feature of her face. The mere external features of her face are sufficient cause of my experience, external features which, due to a mixture of my biology and my socialisation, are experienced as though they embodied sadness.

3. Turing (1950).

4. To be more precise, the Turing test establishes thought or understanding only in terms of *external* behaviour, whereas functional understanding also incorporates the behaviour of the parts inside the system.

5. The term 'zombie' was first coined by Kirk (1974) but the zombie argument is most associated with David Chalmers (1996).

6. James (1890: ch. XIII).

7. The 'consciousness collapses the wave function' interpretation has recently received rigorous treatment from David Chalmers and Kelvin McQueen (Chalmers and McQueen 2022), although they do not necessarily endorse the view.

8. For some recent defences of the Many Worlds interpretation, see Wallace (2012) and Carroll (2019).

9. Ney & Albert (2013) is a good collection of essays covering some of the options.

10. There are a couple of alternative ways of interpreting the pilot wave view: (A) David Albert's (1996) version where there is just the wave function and a single

particle; (B) the view that only three-dimensional space is real and the wave function has more of the status of a law of nature than a concrete physical entity (Dürr, Goldstein, & Zanghì 1997; Goldstein & Teufel 2001).

11. Maudlin (2019) is a great introduction to three of the main interpretations of quantum mechanics. Some of the essays in Cushing et al. (1996) make tentative attempts to move towards a Bohmian version of quantum field theory.

12. For readers aspiring in this direction, my paper (2023) 'Quantum Mechanics and the Consciousness Constraint' suggests how consciousness research and foundations of quantum mechanics might connect up. One of my biggest regrets is not studying physics when I was younger, and hence not being able to take these ideas further. If there are any millionaire readers out there who would like to fund me to take an extended sabbatical to get a physics degree, please do get in touch.

13. Dürr, Goldstein, & Zanghì (1997) and Goldstein & Teufel (2001) use this consideration to argue that the wave function is more like a law of nature than a concrete entity. What I defend in this chapter is a different way of resolving this difficulty.

14. I first formulated pan-agentialism in the final chapter of my (2019) book *Galileo's Error*, and later in my (2020a) paper, 'Panpsychism and Free Will.' It is further developed in this chapter.

15. https://www.nytimes.com/2023/07/01/science/consciousness-theories.html

16. I took the term 'detection procedures' from Matthias Michel's (2019) very interesting discussion of this challenge in the history of consciousness science.

17. For an excellent survey of the literature on this debate, see Michel (2022).

18. Schwitzgebel (ms). Keith Frankish and I discussed this with Eric on our YouTube Channel/podcast 'Mind Chat': https://www.youtube.com/watch?v=U_Q7F-kOcBE&t=224s.

19. I defend this intuition at greater length in Goff (2013).

20. Baars (1988); Dehaene (2014).

21. Dennett (1994).

22. Global workspace theory and the idea that consciousness is 'fame in the brain' are sometimes distinguished, but the same point would apply to either theory. I group them together here for the sake of ease of exposition.

23. Tononi et al. (2016).

24. Mørch (2019).

25. Carroll (2016) gives a vivid exposition of this kind of position.

26. To be more precise, the behaviour will depart from what is specified by the 'Core theory,' which combines the standard model of particle physics and the weak limit of general relativity.

27. One might want to press that this view *does* violate the predictions of quantum mechanics, which are surely simply the predictions entailed by the physical theory. I am sympathetic to Nancy Cartwright's (1983) view that physical laws involve an implicit *ceteris paribus* clause. In other words, the laws of physics specify what will happen *all things being equal*. Even if one rejects this semantic claim, the deeper point is that the full metaphysical theory I am proposing

allows for deviations from the predictions standardly associated with quantum mechanics (whether or not we want to insist that those predictions are the only predictions it is permissible to call 'the predictions of quantum mechanics').

28. Cobb (2020) was the book that most persuaded me of this.

29. https://www.nobelprize.org/uploads/2018/06/laughlin-lecture.pdf

30. Anderson (1972).

31. Picard & Sandi (2021).

32. Mitchell (2018); Sharma et al. (2022).

33. Cutter and Crummett (forthcoming) is an excellent account of the mystery of psycho-physical harmony, but formulated as an argument for the existence of God. Saad (2019) instead explains psycho-physical harmony in terms of a fundamental teleological law. Psycho-physical harmony, or something like it, has also been discussed in Latham (2000); Langsam (2011); Pautz (ms, 2010, 2013, 2020); Mørch (2017, 2020); Goff (2018a); Chalmers (2018).

34. Of Anglophone philosophers, 60 per cent think that philosophical zombies are logically coherent, which implies that there's no logical implication from our behaviour to our conscious experience: https://survey2020.philpeople.org/survey/results/4930

35. I first introduced Inverted Ian to the world in Goff (2018a). There is a precedent for these kinds of 'mismatch cases' in Pautz (2013).

36. Jackson (1982); Robinson (2019). Mørch (2017, 2020) argues that, although epiphenomenalism is conceivable, we can know *a priori* that if pain does *anything*, then it disposes its bearer to try to avoid it. Langsam (2011) has argued for something similar. I think these philosophers are confusing the *rational* absurdity of Inverted Ian, i.e. the fact that his behaviour is matched with his consciousness in a highly inappropriate way, for a (broadly) *logical* absurdity, i.e. that there is some contradiction or incoherence in the idea of Inverted Ian. One reason to be skeptical of their view is that there certainly is a connection of rational appropriateness between pain and avoidance behaviour, and it would be quite a fluke if that connection just happened to go along with a connection of (broadly) logical entailment. In any case, as Cutter and Crummett (forthcoming) have pointed out, these connections are between different conscious states rather than between consciousness and behaviour, and hence the problem is not avoided even if Mørch and Langsam are correct.

37. Orwell (1968: 125).

38. Brian Cutter and Dustin Crummett (forthcoming) argue for the existence of God on the basis of psycho-physical harmony. Chapter 4 outlines my concerns with the God hypothesis. I have debated God versus pan-agentialism as a solution to the psycho-physical harmony problem with Cutter on the YouTube channel 'The Analytic Christian', and with Crummett on Emerson Green's YouTube channel.

39. It's also possible the Earth would have been populated by total zombies, if the universe had contained non-conscious matter, or conscious particles with no capacity to combine into conscious systems.

40. From the 2020 Philpapers survey: https://survey2020.philpeople.org/survey/results/4838

41. Strictly speaking, compatibilism is simply the view that determinism is *compatible* with free agency, and therefore compatibilism doesn't entail that our choices are causally determined. However, compatibilists do tend to think our choices are determined (or at least that the only exceptions to determinism are random quantum events).

42. Strawson (1994).

43. On some interpretations of quantum mechanics, there are no truly random happenings. But we can assume here, for the sake of discussion, an interpretation on which is genuine randomness.

44. Wittgenstein (1953/2010: ch. 1, v. 1).

45. Could it be that meaning zombies are coherent but Inverted Ian is incoherent? This combination of views is tempting, but hard to make sense of when you think about it carefully. We would have to suppose that there is some subtle conceptual connection between human functional understanding and human experiential understanding, which renders it *non*-contradictory to conceive of the former in the absence of the latter but contradictory to conceive of certain unusual combinations of the former and the latter. Maybe it's my limited imagination, but I can't think what that conceptual connection might be. I suspect that the reason it's more tempting to think Inverted Ian is incoherent than a regular zombie, is simply that there's something so obviously absurd about Inverted Ian. However, what's absurd about Inverted Ian is not that he's incoherent but that he's behaving so profoundly irrationally. There's nothing incoherent about the idea of a creature that behaves incredibly irrationally.

46. See Bayne & Montague (2011) for a good collection of essays on *cognitive phenomenology*, defined as a kind of experience essentially tied to thought and understanding, and Kriegel (2013) for a good collection of essays on *phenomenal intentionality*, defined as mental representation that is grounded in consciousness.

47. See, for example, the essays by Carruthers & Veillet, Printz, and Tye & Wright in Bayne and Montague (2011).

Chapter 4: Why the Omni-God Probably Doesn't Exist

1. This horrific case was introduced into discussions of the problem of evil by Bruce Russell (1989).

2. We discussed compatibilism in the 'Digging Deeper' section of Chapter 3. If compatibilism is true, then God could have created a world in which everyone is determined to freely do good.

3. Rasmussen & Leon (2020: 210–11).

4. Swinburne (2004: Ch. 11).
5. Swinburne (2004: 257).
6. Swinburne (2004: 264).
7. I'd like to thank Emerson Green for drawing my attention to this passage with a Tweet.
8. The Kalām Argument was originally formulated by the 11th-century Persian Muslim philosopher Al-Ghazali, but it is most associated in contemporary philosophy with the Christian philosopher William Lane Craig (1991).
9. Contingency arguments for God are most associated historically with the 17th-century philosopher, Gottfried Wilhelm Leibniz. See Pruss (2009) for a good discussion of such arguments.
10. I mean 'prior' here in an explanatory rather than a temporal sense.
11. Rasmussen (2019).
12. This example is from Swinburne (2004: 97).
13. Rasmussen argues that we know the Ultimate Foundation has *some* value—it is inherently valuable to have the capacity to create things of value, for example. And if the Ultimate Foundation does not contain arbitrary limits, then it must be maximally perfect, given that it has some value. In response to this, I would make the same argument I made in the main text. Even granting that the postulation of a perfect being cries out *less* for explanation than the postulation of a being with a certain limited degree of perfection, it still calls for explanation *to an extent*. At the very least, we can ask why a perfect being exists as opposed to nothing at all.
14. Plantinga (1985: 35).
15. Romans 11:33–4.
16. Plantinga (1974: 10).
17. These analogies are from Alston (1996).
18. Wykstra (1984).
19. Rowe (1979).
20. McBrayer (2010). If we were dealing with an inductive inference, then I think these concerns about whether or not our sample is representative would be pertinent. However, as we'll discover in the next section, I don't think the argument from evil should be construed as an inductive inference.
21. If there are strong arguments for the existence of God, this would count against the Cosmic Sin Intuition, because if there really is a God who is perfectly good, then, by definition of Her perfect goodness, God must have some reason why She allows the suffering we see. In my view, the best argument for God is the fine-tuning argument. However, as we'll see in Chapter 5, there are other ways of accounting for the cosmic purpose evidenced by the fine-tuning. In the absence of a good argument for the existence of the Omni-God, it is rational to trust the Cosmic Sin Intuition and infer that the Omni-God doesn't exist.

22. John Mackie (1955, 1982) is a prominent 20th-century exponent of the logical problem of evil. Alvin Plantinga (1974, 1977) is generally credited with refuting the logical problem of evil, and decisively moving the debate on to the evidential problem of evil.

Chapter 5: Cosmic Purpose without God

1. There are of course some exceptions. The following books had a huge influence on me: Leslie (1989); Nagel 2012; Mulgan (2015).

2. Law (2010).

3. I don't, however, agree with Law that his 'evil God challenge' is an effective argument against the Omni-God. Believers in the Omni-God could agree with Law that Good God and Evil God are equal as regards the problem of evil and the problem of good, but nonetheless think there are other considerations in favour of Good God over Evil God. Having said that, I also think—as I hope I made clear in Chapter 4—that the more familiar problem of evil is more than sufficient to rule out the Omni-God.

4. Draper (2017/2022).

5. I'm simplifying a little, as it depends on the level at which the relevant structure is required, but we can imagine that our robot realizes the relevant structure at the right level.

6. If you're a panpsychist, doesn't that mean that consciousness is everywhere, and hence that the silicon robot would be conscious? Not quite. For a panpsychist, all physical systems involve consciousness *at some level*, but not necessarily at the macro level. It's an open question whether the silicon robot would have consciousness associated with its brain or merely with its particles. The answer depends on the ongoing empirical question of when mental combination occurs. I tentatively defended the integrated information theory as an answer to this in Chapter 3. If this hypothesis turns out to be correct, then whether a macro-level system in the brain of the silicon robot is conscious will depend on whether there is more integrated information in that system than in its parts.

7. Bostrom (2003). To be more precise, the paper argues that at least one of the following propositions is true: (1) the human species is very likely to become extinct before reaching a 'posthuman' stage; (2) any posthuman civilization is extremely unlikely to run a significant number of simulations of their evolutionary history (or variations thereof); (3) we are almost certainly living in a computer simulation.

8. Alan Guth articulates this in a discussion on the 'Closer to Truth' series, available on YouTube: https://www.youtube.com/watch?v=5ZtRfACbygY&t=2s

9. Tim Mulgan (2015) defends an Omni-God who doesn't care about us. Mulgan's book is excellent, and a huge inspiration for what I am doing here; Mulgan is also defending cosmic purpose in the absence of the Omni-God. However,

Mulgan's book is quite challenging, and I think the accessibility of this book would be sacrificed by engaging it in the main text. I critique Mulgan's view in my review of his book (Goff 2022a). Very roughly, I don't find it plausible that there could be a perfectly good being who doesn't care about the intense suffering of creatures, no matter how unimpressive those creature are from the Omni-God's perspective.

10. The facts about value secured here are not necessarily the primitive value facts committed to by the value fundamentalism I expressed sympathy for in Chapter 1 (that would require further argument); nonetheless, the threat of value nihilism is dealt with. One might worry the vacuousness charge returns in the following way. If we are not starting from a commitment to certain things being of objective value, then anything could turn out to be of objective value, e.g. the most objectively valuable universe could turn out to be one filled with only hydrogen. And if anything could turn out to be objectively valuable, then the postulation of a value-responsive cosmic designer predicts nothing. To get around this problem, we can build into the hypothesis being supported that our idealized value intuitions roughly track the truth.

11. The Evil Designer and the designer of the simulation hypothesis are also value-responsive designers, at least as I'm thinking of them. The Evil Designer has fine-tuned the universe to create things of value—conscious organisms—because there's a great evil in destroying or hurting creatures with inherent moral worth; similarly, the designer of the simulation hypothesis has deliberately created valuable living organisms because it's inherently interesting to research such things. If we entirely drop the implicit assumption that these designers are responding to objective value considerations—e.g. if the Evil Designer is so evil she wants to create randomly rather than purposively, or if we are considering possible simulators who are not responding in any way to objective value facts—then these hypotheses collapse into the vacuousness that the Amoral Designer Hypothesis suffers from (unless the hypothesis is further qualified). If the cosmic designer is not responding to objective value, then there is no constraint on what desires she might have, and thus a generic design hypothesis predicts nothing.

12. Nagel (1974).
13. Murphy & Nagel (2002).
14. Nagel (2012).
15. https://twitter.com/sapinker/status/258350644979695616
16. Leiter & Weisberg (2012).
17. In talking of teleological 'laws of nature', I am not meaning to take a stand on what laws of nature are. Some philosophers (Armstrong 1983) take laws of nature to be fundamental components of reality; others take them to be grounded in the powers or capacities of objects (Bird 2005); others (Lewis 1983) take laws of nature to be simply summaries of the regular ways in which entities are arranged across space and time.

18. Hawthorne & Nolan (2006).
19. Leslie (1989).
20. Leslie's view is a form of *pure axiarchism*, according to which the entire universe exists because it is good that it exists. We are here considering a *partial* form of axiarchism, on which a drive towards the good plays *some* role in shaping the development of the universe, together with other, non-value-based factors, i.e. the laws of physics. Whereas Leslie ambitiously aims to explain *why the universe exists*, in terms of axiarchism, a more modest axiarchism would simply aim to embed axiarchic forces within the broader scheme of causal factors influencing the development of the universe.
21. Chalmers (2022: ch. 15). This argument also appeared in Chalmers (1995 and 1996).
22. Schneider (2019).

Chapter 6: A Conscious Universe

1. Russell (1927).
2. Physics is arguably not purely mathematical, as physical theories employ the concept of a law of nature, which is not a mathematical concept. We can say that physics is purely *quantitative*, in the sense of involving only mathematical and causal notions.
3. Another option is to hold that fundamental reality is purely mathematical, that the physical universe is made up of numbers, functions, and vectors (Tegmark 2014), or slightly less radically that the physical universe is constituted of causal structure (Mumford 2004). Following Russell, I've argued (Goff 2017: 6.1.2) that the latter views involve vicious regress in their attempts to articulate the nature of reality. Moreover, as I argue in this chapter (especially in the 'Digging Deeper' section), such views cannot account for the reality of consciousness, and so cannot be true accounts of the actual world, as consciousness exists in the actual world.
4. Strawson (2006); Nagel (2012); Coleman (2012); Chalmers (2015); Alter & Nagasawa (2015); Brüntrup & Jaskolla (2016); Goff (2017); Mørch (2019); Roelofs (2019); Seager (2020).
5. Hawking (1988).
6. This was a 2019 blog post called 'Electrons Don't Think': https://backreaction. blogspot.com/2019/01/electrons-dont-think.html
7. In a sense, naturalistic dualists believe consciousness 'arises' from the physical, as the psycho-physical laws ensure that consciousness comes into being when certain physical conditions are in place. We can add that the materialist thinks consciousness arises from the physical facts *alone*, without the need for such extra laws of nature. For a more precise definition of materialism, see chapter 2 of my (2017) book *Consciousness and Fundamental Reality*.

8. The word 'idealism' is also used for views on which fundamental reality consists wholly of mind or consciousness. There is much debate on how to understand the difference between 'idealism' and 'panpsychism'. I tend to think of idealists as denying the reality of the physical world. For example, Donald Hoffman (2019) argues that the physical world is merely a useful 'user interface'. I raise some objections to Hoffman's position in Goff (2021b), in response to Robert Prentner (2021; both essays are reprinted in Goff & Moran 2022). However, clearly views which fall under these labels are natural bedfellows, and I'm not too concerned to fight over whether we're called the Judean People's Front or the People's Front of Judea. There are also 'neutral monist' views, according to which fundamental reality is neither physical nor mental but somehow gives rise to both (Stoljar 2001; Coleman 2016; McClelland 2013; Schneider 2017). The difficulty here is in giving some positive characterization of this 'neutral' element. *Prima facie*, our options seem confined to what we know from science—which seems to lead us to the physical—or what we know from introspection—which seems to lead to the mental. Coleman (2016) takes the neutral element to be *unexperienced qualities*, known from introspection but non-mental.

9. Chalmers (1996); Nida-Rümelin (2009); Weir (forthcoming).

10. Three of our debates are viewable online: Season 1, episode 5 of *Mind Chat* (https://www.youtube.com/watch?v=NkMpFcDwDxM&t=5130s); the 'Mystery of Consciousness' debate in Liverpool, hosted by the *Panpsycast* podcast and the *Unbelievable?* podcast (https://www.youtube.com/watch?v=WbYs-diyD-I&t=2377s); the Royal Institute of Philosophy annual debate 'The Science and Philosophy of Self-Consciousness' (https://www.youtube.com/watch?v=vjvLQ7GKxBE).

11. Seth (2021).

12. Goff (2006, 2009).

13. Goff (forthcoming).

14. I suppose one could think that conscious experiences are identical to strongly emergent processes that are nonetheless purely physical, in the sense that their nature can be fully explicated in purely quantitative terms. However, once the central argument for materialism (rooted in the supposed empirical evidence for micro-reductionism, aka 'causal closure'; Papineau 2001) has been empirically refuted, there will be nothing to counterbalance the philosophical case against physicalism about consciousness sketched in the 'Digging Deeper' section of this chapter.

15. Zorgan (2073).

16. Goff (2017: ch. 9, 2018b, 2020b, forthcoming).

17. Goff (2018b). I also published a popular version of this argument with *Aeon* magazine: https://aeon.co/essays/cosmopsychism-explains-why-the-universe-is-fine-tuned-for-life. Credit where credit is due: excellent amateur philosopher Justin Gaudry came up with the idea of explaining fine-tuning in terms of

cosmopsychism in a blog post in 2008 (https://panexperientialism.blogspot. com/2008/08/goldilocks-enigma.html?m=1).

18. I have previously called this view 'agentive cosmopsychism', but I figured this name would be a bit confusing given I'm also defending a theory called 'pan-agentialism' in this book.

19. Humeans (Lewis 1983) about laws of nature argue that there is ultimately nothing driving the predictable behaviour of the universe. But I agree with Galen Strawson (1987, 1989/2014) that it's too much of a 'cosmic coincidence' to think the universe continues to behave in such a regular manner from moment to moment in the absence of something to ensure this. In any case, Humeans cannot adopt the view that the world is pure causal structure, at least as it's standardly understood, as this view postulates fundamental causal powers, which Humeans don't believe in. I concede that for Humeans who adopt the view that the universe is pure mathematical structure, there will be no pressure—by the lights of their view—to postulate some causal factor underlying the fundamental regularities of the universe. Still, there would not be consciousness in a purely mathematical universe, and so we know that this view cannot be true of the actual world.

20. Justin Gaudry has suggested that we might account for the foresight of the universe in terms of backwards causation (https://panexperientialism.blogspot. com/2008/08/goldilocks-enigma.html?m=1).

21. I'm getting this figure by adding the philosophers who think zombies are conceiv-- able but not possible to the philosophers who think zombies are metaphysically possible (https://survey2020.philpeople.org/survey/results/4930). I suppose in principle there could be philosophers who think that the qualities of conscious experience are conceptually entailed by the physical facts, but the fact that they are *experienced* is not. However, whereas there are philosophers (McClelland 2013; Coleman 2016) who think the physical facts don't conceptually entail which qualities are involved in our experience but do conceptually entail that they are experienced, I don't know of any philosophers who hold the converse. In any case, I'm assuming the vast majority of the 60 per cent of Anglophone philosophers who think that zombies are conceivable think that neither the qualities of our experience nor the fact that they are experienced is conceptu-ally entailed by the facts of physics.

Chapter 7: Living with Purpose

1. Miller et al. (2019).

2. Byock (2018).

3. For a good recent study, see Andersen et al. (2020). For a good overview of some recent studies, see: https://dana.org/article/psychedelics-weighing-the-healing-power/

4. James (1902).
5. Otto (1917/1923).
6. James (1896).
7. Borg (1997).
8. Hick (1989).
9. Religious fictionalism comes in a variety of strengths, with some denying the literal reality of anything beyond than the facts of empirical science (Scott & Malcolm 2018). I am here exploring a kind of partial fictionalism, which combines a generic commitment to the 'More' with a fictionalist understanding of the culturally specific claims of each religion.
10. This paragraph, and the paragraph on apophatic Christianity above, are taken from Goff (2022b).
11. According to 2021 UK census, the percentage of people declaring they had 'no religion' has soared from 14.8 per cent twenty years ago to 37.2 per cent in 2022 (https://www.theguardian.com/uk-news/2022/nov/29/leicester-and-birmingham-are-uk-first-minority-majority-cities-census-reveals). A 2021 Gallup poll in the US also found religious practice in decline (https://news.gallup.com/poll/1690/religion.aspx).
12. https://www.theosthinktank.co.uk/research/2022/04/21/science-and-religion-moving-away-from-the-shallow-end
13. Piketty (2022: ch. 7).
14. Piketty et al. (2011) argue that there is almost zero correlation between high growth and low top marginal tax rates. This article is a simple overview of their argument that the top marginal rate of taxation could be over 80 per cent (https://cepr.org/voxeu/columns/taxing-1-why-top-tax-rate-could-be-over-80).
15. Piketty (2014: 68, n. 33).
16. I'm inclined to think all property rights are constructed. But, as I discuss in the appendix to this chapter ('P.S. Is Taxation Theft?'), I'm somewhat open to the 'left-libertarian' view that there are natural property rights over one's body and the fruits of one's labour. In any case, we can justify Piketty's economic model on the denial of natural property rights to land and natural resources.
17. Piketty (2020: ch. 17).

P.S. Is Taxation Theft?

1. Nozick (1974); Rothbard (1982).
2. Locke (1689/1988).
3. Steiner (1994); Vallentyne & Steiner (2000); Otsuka (2003).
4. Shrubsole (2019).

Acknowledgements

I'm very lucky to be able to talk about my work with a wide variety of people, both professional and (often brilliant) amateur philosophers. I'm very grateful to all, especially those who disagree with me.

For incredibly helpful comments on drafts, I'd like to thank: Peter Momtchiloff, Mark O'Brien (@disagreeable_I on Twitter), Andrei Buckareff, Robin Le Poidevin, Jack Symes, Simon Goff, and three anonymous reviewers.

For instructive feedback in talks, I'm grateful to all of my students of philosophy of mind and philosophy of religion, as well as the audiences at talks I gave on the meaning zombie problem and psycho-physical harmony at Davidson University, the Jowett Society (Oxford University), Edinburgh Undergraduate Philosophy Society, and the *Consciousness: An Interdisciplinary Perspective* conference (Uehiro Centre, Oxford University).

I owe a great deal to each and every member of the Durham Philosophy Club (aka my PhD students when we go for a beer): Duncan Lee, Elle Takashima, Gaurav Kudtarkar, Jack Symes, and (honorary member) Mark O'Brien.

I'm eternally grateful for physics help from Luke Barnes, Geraint Lewis, Sean Carroll, Barry Loewer, and Graham White, for meta-ethics advice from James Lenman, Bart Streumer, David Faraci, and Chris Cowie, and for invaluable discussions on probability with Paul Draper, Alan Hajek, and Barry Loewer (even though Barry gets cross with me . . .).

This book was massively improved by wise and thorough feedback on all aspects by Peter Momtchiloff.

I'm grateful to Aeon Magazine for allowing me to reprint (a slightly modified version of) my article 'Is Taxation Theft?' as the appendix to this book. https://aeon. co/essays/if-your-pay-is-not-yours-to-keep-then-neither-is-the-tax

Thanks to Peter Momtchiloff and Emma Goff for their input to the cover. Peter suggested Aztec ruins, which made Emma think of the series of photos from which the cover photo was chosen; I then chose the specific photo, and the design team did a great job putting it together.

A huge thank you to Emma Goff for the wonderful illustration of Jesus on toast.

So many people contributed to discussions over the title. I'm sure there were others, but thanks to Peter Momtchiloff, my grandmother-in-law Ann Wood, and Aldo Capalletti for (independently) coming up with the title we eventually went for.

For ongoing stimulating discussions, I'm thankful to John Houghton and Simon Goff.

I'm very lucky to have tremendous love and support from my parents Marie and Tony, and my siblings Helen, Clare, and Simon.

As always, my biggest debt is to Emma Goff for her unconditional love and support, as well as for our many in depth discussions of these issues, most of which were held against the backdrop of a screaming child or two.

Bibliography

Adams, Douglas (2002) *The Salmon of Doubt*, William Heinemann Ltd.

Albert, David Z. (1996) 'Elementary Quantum Metaphysics', in J. T. Cushing, A. Fine, & S. Goldstein (eds.), *Bohmian Mechanics and Quantum Mechanics: An Appraisal*, Kluwer, 277–84.

Alexander, Samuel (1920) *Space, Time, and Deity*, Macmillan & Co Ltd.

Alston, William P. (1996) 'Some (Temporarily) Final Thoughts on the Evidential Arguments from Evil', in Daniel Howard-Snyder (ed.), *The Evidential Argument from Evil*, Indiana University Press, 311–32.

Alter, Torin & Yujin Nagasawa (eds.) (2015) *Consciousness in the Physical World: Essays on Russellian Monism*, Oxford University Press.

Kristoffer A. A. Andersen, Robin Carhart-Harris, David J. Nutt, David Erritzoe (2020) 'Therapeutic Effects of Classic Serotonergic Psychedelics: A Systematic Review of Modern-era Clinical Studies', *Acta Psychiatrica Scandinavica*, 143(2), 101–18.

Anderson, Philip W. (1972) 'More Is Different: Broken Symmetry and the Nature of the Hierarchical Structure of Science', *Science* 177(4047), 393–6.

Armstrong, David, M. (1983) *What Is a Law of Nature?*, Cambridge University Press.

Ayer, A. J. (1936) *Language, Truth and Logic*, Victor Gollancz Ltd.

Baars, Bernard J. (1988) *A Cognitive Theory of Consciousness*, Cambridge University Press.

Barnes, Luke (2020) 'A Reasonable Little Question: A Formulation of the Fine-Tuning Argument', *Ergo* 6, 1220–57.

Bayne, Tim & Michelle Montague (eds.) (2011) *Cognitive Phenomenology*, Oxford University Press.

Benaceraff, Paul (1973) 'Mathematical Truth', *Journal of Philosophy* 70(19), 661–79.

Benatar, David (2006) *Better Never to Have Been*, Oxford University Press.

Benatar, David (2017a) *The Human Predicament*, Oxford University Press.

Benatar, David (2017b) 'Kids? Just Say No.' *Aeon*. https://aeon.co/essays/having-children-is-not-life-affirming-its-immoral

Bird, Alexander (2005) 'The Dispositionalist Conception of Laws', *Foundations of Science* 10(4), 353–70.

Borg, Marcus (1997) *The God We Never Knew: Beyond Dogmatic Religion to a More Authentic Contemporary Faith*, Harper Collins.

Bostrom, Nick (2003) 'Are You Living in a Computer Simulation?' *Philosophical Quarterly* 53(211), 243–55.

Bradley, Darren J. (2009) 'Multiple Universes and Observation Selection Effects', *American Philosophical Quarterly* 46, 61–72.

Brüntrup, Godehard & Ludwig Jaskolla (eds.) (2016) *Panpsychism: Contemporary Perspectives*, Oxford University Press.

Byock, Ira (2018) 'Taking Psychedelics Seriously', *Journal of Palliative Medicine* 21(4), 417–21.

Camus, Albert (1942) *Le Mythe de Sisyphe*, Éditions Gallimard.

Carroll, Sean (2016) *The Big Picture: On the Origins of Life, Meaning, and the Universe Itself*, Dutton.

Carroll, Sean (2019) *Something Deeply Hidden: Quantum Worlds and the Emergence of Spacetime*, Dutton.

Cartwright, Nancy (1983) *How the Laws of Physics Lie*, Oxford University Press.

Chalmers, David, J. (1995) 'Facing Up to the Problem of Consciousness', *Journal of Consciousness Studies* 2(3), 200–19.

Chalmers, David, J. (1996) *The Conscious Mind: In Search of a Fundamental Theory*, Oxford University Press.

Chalmers David, J. (2015) 'Panpsychism and Panprotopsychism', in Alter & Nagasawa 2015.

Chalmers, David, J. (2018) 'The Meta-Problem of Consciousness', *Journal of Consciousness Studies* 25(9–10), 6–61.

Chalmers, David, J. (2022) *Reality+: Virtual Reality and the Problems of Philosophy*, Allen Lane.

Chalmers, David, J. & McQueen, Kelvin (2022) 'Consciousness and the Collapse of the Wave Function', in Shan Gao (ed.), *Consciousness and Quantum Mechanics*, New York: Oxford University Press.

Cobb, Matthew (2020) *The Idea of the Brain: A History*, Profile Books.

Coleman, Sam (2012) 'Mental Chemistry: Combination for Panpsychists', *Dialectica* 66(1), 137–66.

Coleman, Sam (2016) 'Panpsychism and Neutral Monism: How to Make Up One's Mind', in Brüntrup & Jaskolla 2016.

Collins, Robin (2009) 'The Teleological Argument: An Exploration of the Fine-Tuning of the Universe', in William Lane Craig & J. P. Moreland (eds.), *The Routledge Companion to Natural Theology*, Routledge, 202–81.

Colyvan M., J. L. Garfield, & G. Priest (2005) 'Problems with the Argument from Fine-Tuning', *Synthese* 145, 39.

Craig, William Lane (1991) 'The Existence of God and the Beginning of the Universe', *Truth* 3.

Cushing, James T., Arthur Fine, & Sheldon Goldstein (eds.) (1996) *Bohmian Mechanics and Quantum Theory: An Appraisal*, vol. 184: *Boston Studies in the Philosophy of Science*, Kluwer Academic Publishers.

Cutter, Brian & Dustin Crummett (forthcoming) 'Psychophysical Harmony: A New Argument for Theism', *Oxford Studies in Philosophy of Religion*, Oxford University Press.

Darwin, Charles (1859) 'Letter No. 2532, Charles Darwin to John Lubbock', *Darwin Correspondence Project*, 22 November.

Davies, Paul C.W. (2006) *The Goldilocks Enigma: Why Is the Universe Just Right for Life?*, Allen Lane.

Dawkins, Richard (1986) *The Blind Watchmaker*, Norton & Co.

Dehaene, Stanislas (2014) *Consciousness and the Brain: Deciphering How the Brain Codes Our Thoughts*, Viking Adult.

Dennett, Daniel (1994) 'Real Consciousness', in Antti Revonsuo & Matti Kamppinen (eds.), *Consciousness in Philosophy and Cognitive Neuroscience*, Lawrence Erlbaum Associates, 55–63.

Draper, Kai, Paul Draper, & Joel Pust (2007) 'Probabilistic Arguments for Multiple Universes', *Pacific Philosophical Quarterly* 88(3), 288–307.

Draper, Paul (2020) 'In Defense of the Requirement of Total Evidence', *Philosophy of Science* 87(1), 179–90.

Draper, Paul (2017/2022) 'Atheism and Agnosticism', *Stanford Encyclopedia of Philosophy*, Stanford University Press (original entry published in 2017; revised in 2022).

Dürr, D., Goldstein, S., & Zanghì, N. (1997) 'Bohmian Mechanics and the Meaning of the Wave Function', in R. S. Cohen, M. Horne, & J. Stachel (eds.), *Experimental Metaphysics—Quantum Mechanical Studies for Abner Shimony*, vol. 1: *Boston Studies in the Philosophy of Science*, Kluwer Academic Publishers, 193.

Frankish, Keith (2016) 'Illusionism as a Theory of Consciousness', *Journal of Consciousness Studies* 23(11–12), 11–39.

Freud, Sigmund (1927/8) *The Future of Illusion*, trans. W. D. Robson-Scott, Hogarth Press.

Goff, Philip (2006) 'Experiences Don't Sum', *Journal of Consciousness Studies* 13(10–11), 53–61.

Goff, Philip (2009) 'Why Panpsychism Doesn't Help Us Explain Consciousness', *Dialectica* 63(3), 289–311.

Goff, Philip (2013) 'Orthodox Property Dualism + Linguistic Theory of Vagueness = Panpsychism', in R. Brown (ed.), *Consciousness Inside and Out: Phenomenology, Neuroscience, and the Nature of Experience*, Springer, 75–91.

Goff, Philip (2017) *Consciousness and Fundamental Reality*, Oxford University Press.

Goff, Philip (2018a) 'Conscious Thought and the Cognitive Fine-Tuning Problem', *Philosophical Quarterly* 68(270), 98–122.

Goff, Philip (2018b) 'Did the Universe Design Itself?' *International Journal for Philosophy of Religion* 85, 99–122.

Goff, Philip (2019) *Galileo's Error: Foundations for a New Science of Consciousness*, Pantheon.

Goff, Philip (2020a) 'Panpsychism and Free Will', *Proceedings of the Aristotelian Society* 120(2), 123–44.

Goff, Philip (2020b) 'Cosmopsychism, Micropsychism and the Grounding Relation', in W. Seager (ed.), *The Routledge Handbook of Panpsychism*, Routledge, 144–56.

Goff, Philip (2021a) 'Our Improbable Existence Is No Evidence for a Multiverse', *Scientific American*. https://www.scientificamerican.com/article/our-improbable-existence-is-no-evidence-for-a-multiverse/

Goff, Philip (2021b) 'Putting Consciousness First', *Journal of Consciousness Studies* 28(9–10), 289–328.

Goff, Philip (2022a) 'Review of Tim Mulgan's *Purpose in the Universe: The Moral and Metaphysical Case for Ananthropic Purposivism*, by Tim Mulgan', *International Journal for Philosophy of Religion* 92(3), 177–81.

Goff, Philip (2022b) 'Why Religion without Belief Can Still Make Perfect Sense', *Psyche*. https://psyche.co/ideas/why-religion-without-belief-can-still-make-perfect-sense

Goff, Philip (2023) 'Quantum Mechanics and the Consciousness Constraint', in S. Gao (ed.), *Quantum Mechanics and Consciousness*, Oxford University Press.

Goff, Philip (forthcoming) 'How Exactly Does Panpsychism Explain Consciousness?', *Journal of Consciousness Studies*.

Goff, Philip & Alex Moran (2022) *Is Consciousness Everywhere? Essays on Panpsychism*, Academic Imprint.

Goldman, Alan H. (2009) *Reasons from Within: Desires and Values*, Oxford University Press.

Goldstein, Sheldon & Stefan Teufel (2001) 'Quantum Spacetime without Observers: Ontological Clarity and the Conceptual Foundations of Quantum Gravity', in C. Callender & N. Huggett (eds.), *Physics Meets Philosophy at the Planck Scale*, Cambridge University Press.

Greene, Brian (2011) *The Hidden Reality: Parallel Universes and the Deep Laws of the Cosmos*, Vintage.

Guth, Alan H. (1981) 'Inflationary Universe: A Possible Solution to the Horizon and Flatness Problems', *Physical Review D* 23(2), 347.

Guth, Alan H. (2000) 'Inflation and Eternal Inflation', *Physics Reports* 333, 555–74.

Hacking, Ian (1987) 'The Inverse Gambler's Fallacy: The Argument from Design. The Anthropic Principle Applied to Wheeler Universes', *Mind* 96(383), 331–40.

Hájek, Alan (2019) 'Interpretations of Probability', *The Stanford Encyclopedia of Philosophy*, Stanford.

Hawking, Stephen (1988) *A Brief History of Time: From the Big Bang to Black Holes*, Bantam.

Hawthorne, John & Daniel Nolan (2006) 'What Would Teleological Causation Be?', in John Hawthorne (ed.), *Metaphysical Essays*, Oxford University Press, 265–84.

Hick, John (1989) *An Interpretation of Religion: Human Responses to the Transcendent*, Yale University Press.

Hoffman, Donald (2019) *The Case Against Reality: How Evolution Hid the Truth from Our Eyes*, Allen Lane.

Hume, David (1739/2000) *A Treatise of Human Nature*, ed. David Fate Norton & Mary J. Norton, Oxford University Press.

Hume, David (1753/1882/1964) 'Of National Characters', in T. H. Green and T. H. Grose (eds.), *The Philosophical Works* (London, 1882; repr. Darmstadt, 1964), 253.

Jackson, F. (1982) 'Epiphenomenal Qualia', *The Philosophical Quarterly* 32(127), 127–36.

James, William (1896) 'The Will to Believe', *The New World* 5, 327–47.

James, William (1890) *Principles of Psychology*, Henry Holt & Co.

James, William (1902) *The Varieties of Religious Experience: A Study in Human Nature*, Longmans, Green & Co.

Kammerer, François (2022) 'How can you be so sure? Illusionism and the obviousness of phenomenal consciousness', *Philosophical Studies* 179(9), 2845–67.

Kirk, Robert (1974) 'Sentience and Behaviour', *Mind* 83: 329, 43–60.

Kriegel, Uriah (ed.) (2013) *Phenomenal Intentionality*, Oxford University Press.

Landsman, Klaas (2016) 'The Fine-Tuning Argument: Exploring the Improbability of Our Own Existence', in K. Landsman and E. van Wolde (eds.), *The Challenge of Chance* Springer, 111–29.

Langsam, Harold (2011) *The Wonder of Consciousness: Understanding the Mind Through Philosophical Reflection*, MIT Press.

Latham, Noa (2000) 'Chalmers on the Addition of Consciousness to the Physical World', *Philosophical Studies* 98(1), 71–97.

Law, Stephen (2010) 'The Evil-God Challenge', *Religious Studies* 46(3), 353–73.

Leiter, Brian & Weisberg, Michael (2012) 'Do you only have a brain? On Thomas Nagel', *The Nation* https://www.thenation.com/article/archive/do-you-only-have-brain-thomas-nagel/

Lenman, James & Matthew Lutz (2006/2018) 'Moral Naturalism', *Stanford Encyclopedia of Philosophy*, Stanford University Press.

Leslie, John (1989) *Universes*, Routledge.

Lewis, David (1983) 'New Work for a Theory of Universals', *Australasian Journal of Philosophy*, 61(4), 343–77.

F. Geraint Lewis & Barnes, Luke (2016) *A Fortunate Universe: Life in a Finely-Tuned Cosmos*, Cambridge University Press.

Locke, John (1689/1988) *Two Treatises of Government*, ed. Peter Laslett, Cambridge University Press.

Mackie, J. L. (1955) 'Evil and Omnipotence', *Mind* 64(254), 200–12.

Mackie, J. L. (1982) *The Miracle of Theism*, Oxford University Press.

Marx, Karl (1843/2009) *Critique of Hegel's 'Philosophy of the Right'*, ed. Joseph O'Malley, Cambridge University Press.

Maudlin, Tim (2019) *Philosophy of Physics: Quantum Theory*, Princeton University Press.

McBrayer, Justin, P. (2010) 'Skeptical Theism', *Philosophy Compass* 5(7), 516–623.

McClelland, Thomas (2013) 'The Neo-Russellian Ignorance Hypothesis: A Hybrid Account of Phenomenal Consciousness', *Journal of Consciousness Studies* 20(3–4), 125–51.

McGrew, Timothy, Lydia McGrew, & Eric Vestrup (2001) 'Probabilities and the Fine-Tuning Argument: A Sceptical View', *Mind* 110(449), 1027–38.

Michel, Matthias (2019) 'Consciousness Science Underdetermined: A Short History of Endless Debates', *Ergo* 6, 28.

Michel, Matthias (2022) 'Consciousness and the Prefrontal Cortex: A Review', *Journal of Consciousness Studies* 29(7–8), 115–57.

Miller, Melanie J., Juan Albarracin-Jordan, Christine Moore, & José M. Capriles (2019) 'Chemical Evidence for the Use of Multiple Psychotropic Plants in a 1,000-Year-Old Ritual Bundle from South America', *Proceedings of the National Academy of Sciences of the United States of America* 116(23), 11207–12.

Mitchell Kevin, J. (2018) 'Does Neuroscience Leave Room for Free Will?', *Trends in Neuroscience* 41(9), 573–6.

Moore, G. E. (1903/1993) *Principia Ethica*, Cambridge University Press, 1903; revised edition with 'Preface to the Second Edition' and other papers, ed. T. Baldwin, Cambridge University Press, 1993 (page references are first to the original and then to the revised edition).

Mørch, Hedda Hassel (2017) 'The Evolutionary Argument for Phenomenal Powers', *Philosophical Perspectives* 31(1), 293–316.

Mørch, Hedda Hassel (2019) 'Is the Integrated Information Theory of Consciousness Compatible with Russellian Panpsychism?', *Erkenntnis* 84(5), 1065–85.

Mørch, Hedda Hassel (2020) 'The Phenomenal Powers View and the Meta-Problem of Consciousness', *Journal of Consciousness Studies* 27(5–6), 131–42.

Mulgan, Tim (2015) *Purpose in the Universe: The Moral and Metaphysical Case for Ananthropocentric Purposivism*, Oxford University Press.

Mumford, Stephen (2004) *Laws in Nature*, Routledge.

Murphy, Liam & Thomas Nagel (2002) *The Myth of Ownership: Taxes and Justice*, Oxford University Press.

Nagel, Thomas (1974) 'What Is It Like to Be a Bat?' *The Philosophical Review* 83(4), 435–50.

Nagel, Thomas (2012) *Mind and Cosmos: Why the Materialist, Neo-Darwinian Conception of Nature is Almost Certainly False*, Oxford University Press.

Ney, Alyssa & David Z. Albert (2013) *The Wave Function: Essays on the Metaphysics of Quantum Mechanics*, Oxford University Press.

Nida-Rümelin, Martine (2009) 'An Argument from Transtemporal Identity for Subject-Body Dualism', in Robert C. Koons & George Bealer (eds.), *The Waning of Materialism: New Essays*, Oxford University Press.

Nietzsche, Friedrich (1881/2001) *The Gay Science*, ed. Bernard Williams, trans. Josefine Nauckhoff & Adrian Del Caro, Cambridge University Press.

Nozick, Robert (1974) *Anarchy, State, and Utopia*, Basic Books.

Olson, Jonas (2014) *Moral Error Theory: History, Critique, Defence*, Oxford University Press.

Orwell, George (1968) *George Orwell: In Front of Your Nose, 1945–1950*, vol. 4: *The Collected Essays, Journalism and Letters*, Harcourt Brace Jovanovich.

Otsuka, Michael (2003) *Libertarianism without Inequality*, Clarendon Press.

Otto, Rudolph (1917/1923) *The Idea of the Holy*, Oxford University Press.

Paley, William (1802) *Natural Theology: Or Evidences of the Existence of the Deity Collected From the Appearances of Nature*, R. Faulder.

Papineau, David (2001) 'The Rise of Physicalism', in Carl Gillett & Barry Loewer (eds.), *Physicalism and its Discontents*, Cambridge University Press, 3–36.

Parfit, Derek (1998) 'Why Anything? Why This?', *London Review of Books*, 22 January, 24–7.

Pautz, Adam (ms) 'A Dilemma for Russellian Monists about Consciousness'. https://philpapers.org/rec/PAUCRM

Pautz, Adam (2010) 'Do Theories of Consciousness Rest on a Mistake?' *Philosophical Issues* 20(1), 333–67.

Pautz, Adam (2013) 'Does Phenomenology Ground Mental Content?', in Uriah Kriegel (ed.), *Phenomenal Intentionality*, Oxford University Press.

Pautz, A. (2020) 'Consciousness and Coincidence: Comments on Chalmers', *Journal of Consciousness Studies* 5–6, 143–55.

Picard, Martin & Carmen Sandi (2021) 'The Social Nature of Mitochondria: Implications for Human Health', *Neuroscience & Biobehavioral Reviews* 120, 595–610.

Pigden, C. R. (2009) 'If Not Non-Cognitivism, Then What?', in C. R. Pigden (ed.), *Hume on Motivation and Virtue*, Philosophers in Depth, Palgrave Macmillan.

Piketty, Thomas, Emmanuel Saez, & Stefanie Stantcheva (2011) 'Optimal Taxation of Top Labor Incomes: A Tale of Three Elasticities', *CEPR Discussion Paper* 8675, December.

Piketty, Thomas (2014) *Capital in the Twenty-First Century*, trans. Arthur Goldhammer, Belknap Press of Harvard University Press.

Piketty, Thomas (2020) *Capital and Ideology*, trans. Arthur Goldhammer, Belknap Press of Harvard University Press.

Piketty, Thomas (2022) *A Brief History of Equality*, trans. Steven Rendall, Belknap Press of Harvard University Press.

Plantinga, Alvin (1974) *The Nature of Necessity*, Clarendon Press.

Plantinga, Alvin (1977) *God, Freedom, and Evil*, Eerdmans.

Plantinga, Alvin (1985) 'Self-Profile', in James E. Tomberlin & Peter van Inwagen (eds.), *Alvin Plantinga*, D. Reidel, 3–97.

Prentner, Robert (2021) 'Dr Goff, Tear Down This Wall! The Interface Theory of Perception and the Science of Consciousness', *Journal of Consciousness Studies* 28(9–10), 91–103.

Proudhon, Pierre-Joseph (1840) *Qu'est-ce que la propriété? ou, Recherche sur le principe du Droit et du Gouvernement*, Brocard.

Pruss, Alexander (2009) 'The Leibnizian Cosmological Argument', in William Lane Craig & J. P. Moreland (eds.), *The Blackwell Companion to Natural Theology*, Wiley-Blackwell, 24–100.

Rasmussen, Joshua (2019) *How Reason Can Lead to God: A Philosopher's Bridge to Faith*, InterVarsity Press.

Rasmussen, Joshua & Filipe Leon (2020) *Is God the Best Explanation of Things? A Dialogue*, Palgrave Macmillan.

Rees, Martin (2000) *Just Six Numbers: the Deep Forces that Shape the Universe*, Basic Books.

Robinson, William, S. (2019) *Epiphenomenal Mind: An Integrated Outlook on Sensations, Beliefs, and Pleasure*, Routledge.

Roelofs, Luke (2019) *Combining Minds: How to Think About Composite Subjectivity*, Oxford University Press.

Rothbard, Murray, N. (1982) *The Ethics of Liberty*, Humanities Press.

Rowe, William L. (1979) 'The Problem of Evil and Some Varieties of Atheism', *American Philosophical Quarterly* 16(4), 335–41.

Russell, Bertrand (1927) *The Analysis of Matter*, Kegan Paul.

Russell, Bruce (1989) 'The Persistent Problem of Evil', *Faith and Philosophy* 6(2), 121–39.

Saad, Bradford (2019) 'A Teleological Strategy for Solving the Meta-problem of Consciousness', *Journal of Consciousness Studies* 26 (9–10), 205–216.

Schneider, Susan (2017) 'Idealism, or Something Near Enough', in Kenny Pearce & Tyron Goldschmidt (eds.), *Idealism: New Essays in Metaphysics*, Oxford University Press.

Schneider, Susan (2019) *Artificial You: AI and the Future of Your Mind*, Princeton University Press.

Schroeder, Mark (2007) *Slaves of the Passions*, Oxford University Press.

Schwitzgebel, Eric (ms) 'Borderline Consciousness, When It's Neither Determinately True Nor Determinately False That Experience Is Present'. http://www.faculty.ucr.edu/~eschwitz/SchwitzPapers/BorderlineConsciousness-211220.pdf

Scott, Michael & Finlay Malcolm (2018) 'Religious Fictionalism', *Philosophy Compass* 13(3), 1–11.

Seager, William (2020) *The Routledge Handbook of Panpsychism*, Routledge.

Seth, Anil (2021) *Being You: A New Science of Consciousness*, Faber.

Sharma, Abhishek, Dániel Czégel, Michael Lachmann, Christopher P. Kempes, Sara I. Walker, and Leroy Cronin (2022) 'Assembly Theory Explains and Quantifies the Emergence of Selection and Evolution', *Arxiv*.

Shrubsole, Guy (2019) *Who Owns England? How We Lost Our Green and Pleasant Land, and How to Take It Back*, William Collins.

Sider, Theodore (2011) *Writing the Book of the World*, Oxford University Press.

Smart, J. J. C. (1989) *Our Place in the Universe: A Metaphysical Discussion*, Blackwell.

Smith, Michael (1994) *The Moral Problem*, Blackwell.

Smolin, Lee (1997) *The Life of the Cosmos*, Oxford University Press.

Sober, Elliott (2003) 'The Design Argument', in Neil A. Manson (ed.), *God and Design: The Teleological Argument and Modern Science*, Routledge, 27–54.

Sober, Elliott (2009) 'Absence of Evidence and Evidence of Absence: Evidential Transitivity in Connection with Fossils, Fishing, Fine-Tuning and Firing Squads', *Philosophical Studies* 143(1), 63–90.

Steiner, H. (1994) *An Essay on Rights*, Blackwell Publishers.

Stoljar, Daniel (2001) 'Two Conceptions of the Physical', *Philosophy and Phenomenological Research* 62(2), 253–81.

Strawson, Galen (1987) 'Realism and Causation', *The Philosophical Quarterly* 37(148), 253–77.

Strawson, Galen (1994) 'The Impossibility of Moral Responsibility', *Philosophical Studies* 75, 5–24.

Strawson, Galen (2006) 'Realistic Monism: Why Physicalism Entails Panpsychism', *Journal of Consciousness Studies* 13(10–11), 3–31.

Strawson, Galen (1989/2014) *The Secret Connection: Causation, Realism, and David Hume, Revised Edition*, Oxford University Press (originally published in 1989).

Streumer, Bart (2017) *Unbelievable Errors: An Error Theory about All Normative Judgements*, Oxford University Press.

Susskind, Leonard (2005) *The Cosmic Landscape: String Theory and the Illusion of Intelligent Design*, Back Bay Books.

Swinburne, Richard (2004) *The Existence of God*, 2nd edn, Clarendon Press.

Tegmark, Max (2014) *Our Mathematical Universe: My Quest for the Ultimate Nature of Reality*, Knopf.

Teilhard de Chardin, Pierre (1955) *The Phenomenon of Man*, Éditions du Seuil.

Than, Ker (2018) 'Is Our Universe One of Many?' *Stanford News*, https://news.stanford.edu/2018/09/10/landscape-theory/

Tooley, Michael (2015) 'The Problem of Evil', *The Stanford Encyclopedia of Philosophy*.

Tononi, Giulio, Melanie Boly, Marcello Massimini, & Christof Koch (2016) 'Integrated Information Theory: From Consciousness to Its Physical Substrate', *Nature Reviews Neuroscience* 17, 450–61.

Turing, Alan (1950) 'Computing Machinery and Intelligence', *Mind* 59(236), 433–60.

Vallentyne, Peter & Hillel Steiner (eds.) (2000) *Left-Libertarianism and Its Critics: The Contemporary Debate*, Palgrave.

van Roojen, Mark (2004/2018) 'Cognitivism versus Non-Cognitivism', *Stanford Encyclopedia of Philosophy*, Stanford University Press.

Wallace, David (2012) *The Emergent Multiverse*, Oxford University Press.

Weir, Ralph (forthcoming) *The Mind-Body Problem and Metaphysics*, Routledge.

White, Roger (2000) 'Fine-Tuning and Multiple Universes', *Noûs* 34(2), 260–67.

Williams, Bernard A. O. (1979/1981) 'Internal and External Reasons', *Moral Luck*, Cambridge University Press, 101–13.

Willams, Bernard, A. O. (1989/1995) 'Internal Reasons and the Obscurity of Blame', *Making Sense of Humanity*, Cambridge University Press, 35–45.

Wittgenstein, Ludwig (1953/2010) *Philosophical Investigations*, ed. Arif Ahmed, Cambridge University Press.

Wolchover, Natalie (2018) 'Why the Tiny Weight of Empty Space Is Such a Huge Mystery', *Nautilus*. https://nautil.us/why-the-tiny-weight-of-empty-space-is-such-a-huge-mystery-237188/

Wykstra, Stephen J. (1984) 'The Humean Obstacle to Evidential Arguments from Suffering: On Avoiding the Evils of "Appearance"', *International Journal for Philosophy of Religion* 16(2), 73–93.

Zorgan, E. T. (2073) *A Time Traveller's Guide to the Future*, Interdimensional Inc.

Index

For the benefit of digital users, indexed terms that span two pages (e.g., 52–53) may, on occasion, appear on only one of those pages.